Surveys of Modern Mathematics
Volume III

Application of Elementary Differential Geometry to Influence Analysis

Yat-Sun Poon
Wai-Yin Poon

International Press
www.intlpress.com

高等教育出版社
HIGHER EDUCATION PRESS

Surveys of Modern Mathematics, Volume III
Application of Elementary Differential Geometry to Influence Analysis

Yat-Sun Poon
Department of Mathematics, University of California at Riverside
Wai-Yin Poon
Department of Statistics, The Chinese University of Hong Kong

2010 Mathematics Subject Classification. 53-01, 53A07, 53A30, 53A45, 53B20, 62J05, 97K70.

ISBN: 978-1-57146-252-7

Printed in the United States of America.

16 15 14 13 12   1 2 3 4 5 6 7 8 9

SURVEYS OF MODERN MATHEMATICS

Dedicated to our father Yan Ding Poon
for his 88th birthday

Contents

Preface · xi

Part I Geometry

1 Preliminaries · 3
1.1 Linear algebra · 3
1.1.1 Vectors and matrices · 3
1.1.2 Symmetric bilinear forms · 5
1.1.3 Vector subspaces · 6
1.1.4 Linear maps from $\mathbb{R}^n$ to $\mathbb{R}^n$ · 7
1.1.5 A convention · 9
1.2 Vector calculus · 9
1.2.1 Vector-valued functions and differentials · · · · · · · · · · · · 9
1.2.2 Taylor expansion and extrema · · · · · · · · · · · · · · · · · · · 11
1.2.3 Extrema and Lagrange multiplier theorem · · · · · · · · · · · · 12

2 Euclidean Geometry · 15
2.1 Orthogonal transformations · 15
2.2 Rigid motions · 16
2.3 Translation of vector subspaces · 18
2.4 Conformal transformations · 20
2.5 Orthonormal basis · 20
2.6 Orthogonal projections · 23
2.7 Areas and volumes · 25

3 Geometry of Graphs · 29
3.1 Graphs in Euclidean spaces · 29
3.2 Normal sections · 31
3.3 Cross sections in high dimension · 33

3.4 First fundamental forms · · · 33

4 Curvatures · · · 35
4.1 Normal curvatures · · · 35
4.1.1 Definition · · · 35
4.1.2 Principal curvatures and principal directions · · · 37
4.2 Sectional curvatures · · · 40

5 Transformations and Invariance · · · 43
5.1 Change of coordinates · · · 43
5.2 Non-linear conformal transformations · · · 44
5.3 Invariant curvatures · · · 46

Part II Statistics

6 Discrete Random Variables and Related Concepts · · · 51
6.1 Preliminaries · · · 51
6.2 Discrete random variables · · · 52
6.2.1 Discrete random variables and probability function · · · 52
6.2.2 Relative frequency histogram · · · 55
6.2.3 Cumulative distribution function · · · 55
6.3 Population parameters and sample statistics · · · 56
6.3.1 Population mean and expected value · · · 56
6.3.2 Sample statistic · · · 57
6.3.3 Sample mean · · · 57
6.3.4 Sample and population variances · · · 58
6.4 Mathematical expectations · · · 60
6.5 Maximum likelihood estimation · · · 61
6.6 Maximum likelihood estimation of the probability of a Bernoulli experiment · · · 62

7 Continuous Random Variables and Related Concepts · · · 65
7.1 Continuous random variables · · · 65
7.2 Mathematical expectation for continuous random variables · · · 66
7.3 Mean and variance and their sample estimates · · · 66
7.4 Basic properties of expectations · · · 67
7.5 Normal distribution · · · 68
7.6 Maximum likelihood estimation for continuous variables · · · 72

7.7 Maximum likelihood estimation for the parameters of normal distribution · · · 73
7.8 Sampling distribution · · · 74

8 Bivariate and Multivariate Distribution · · · 77
8.1 Bivariate distribution for discrete random variables · · · 77
8.1.1 Joint probability function · · · 77
8.1.2 Marginal probability function · · · 78
8.1.3 Conditional probability function · · · 79
8.2 Bivariate distribution for continuous random variables · · · 80
8.3 Mathematical expectations · · · 80
8.3.1 Mathematical expectations for the functions of two random variables · · · 80
8.4 Covariance and correlation · · · 82
8.4.1 Sample covariance and correlation · · · 82
8.4.2 Population covariance and correlation · · · 84
8.4.3 Conditional expectations · · · 85
8.5 Bivariate normal distribution · · · 87
8.6 Independence · · · 88
8.7 Multivariate distribution · · · 89

9 Simple Linear Regression · · · 93
9.1 The model · · · 93
9.2 The least squares estimation · · · 95
9.3 The maximum likelihood estimation of regression parameters · · · 98
9.4 Residuals · · · 99
9.5 Coefficient of determination · · · 101
9.6 Weighted least squares estimates · · · 103

10 Topics on Linear Regression Analysis · · · 105
10.1 Multiple regression model · · · 105
10.2 Estimation and interpretation · · · 106
10.3 Influential observations and outliers · · · 110
10.4 Leverage · · · 111
10.5 Cook's distance · · · 113
10.6 Deletion influence, joint influence and masking effect · · · 114
10.7 Derivation of Cook's distances · · · 116
10.7.1 Weighted least squares and Cook's distance · · · 116
10.7.2 Cook's distance – deleting one data point · · · 118

Part III Local Influence Analysis

11 Basic Concepts · · · 123
11.1 Introduction · · · 123
11.2 Perturbation · · · 125
11.3 Likelihood displacement and influence graph · · · 126

12 Measuring Local Influence · · · 129
12.1 Individual influence · · · 130
12.2 Derivation of normal curvature · · · 131
12.3 Case-weight perturbation — an example · · · 133
12.4 Roles of sectional curvature · · · 135
12.5 Joint influence · · · 138

13 Relations Among Various Measures · · · 141
13.1 A bound on influence measures · · · 141
13.2 Individual and overall joint influence · · · 143
13.3 Individual and joint influence measures · · · 146
13.4 Competing eigenvalues · · · 147
13.5 Conclusions · · · 150

14 Conformal Modifications · · · 153
14.1 Modification and invariance · · · 153
14.2 Invariant measures · · · 154
14.3 Benchmarks · · · 155
14.4 Individual's contribution to joint influence — re-visited · · · 157

Appendix A Rank of Hat Matrix · · · 161

Appendix B Ricci Curvature · · · 163

Appendix C Cook's Distance — Deleting Two Data Points · · · 165

Bibliography · · · 167

Index · · · 171

Preface

The two authors of this book have different backgrounds within the mathematical sciences. After fifteen years of collaboration, they find it benefical to introduce their research areas jointly to students in the early stages of their research careers. Mathematics students will learn that mathematical concepts have an immediate impact on real life situations. Statistics students will appreciate that abstract mathematical concepts are accessible, relevant and valuable. In fact, an early draft of this book was used by one of the authors as lecture notes to run one year-long undergraduate research seminar for a mixture of mathematics and statistics undergraduate students in their senior years.

We hope that this book will encourage young scientists to develop an appreciation of inter-disciplinary research at a time when their research career is just beginning to emerge.

Differential geometry is a broad mathematical subject. Its global aspects are often presented only to mathematics majors or graduate students. However, its local aspects are the foundation of global issues and can be made accessible to all students with rigorous, multi-variable calculus training. To study local geometry, one studies graphs. It is a generalization of the freshman discussion on the concavity of a real-valued function of a single variable. By studying the geometry of graphs, one can proceed to learn global differential geometry as a mathematician.

One may also begin to apply the geometry of a graph to analyze functions arising from concrete problems, such as through statistics. Many statistical analyses involve a hypothesized model. Once a model is specified, the data collected are used to estimate the parameters that characterize the model. Further inference is affected by the hypothesized model and the data collected. Therefore, one must assess the influence based on the perturbation of various aspects of the model inputs. Perturbations can be represented by a set of perturbation parameters, and the function of these perturbation parameters becomes a mathematical object of interest. Therefore, one can apply geometric concepts to study this function and hence, deduce information on the perturbation.

Working through this process prompts theoretical, practical and technical issues. In particular, we must develop measures for the influence of individual perturbation parameters. We must also develop measures for the joint influence of any two, and then any groups of perturbation parameters. From single to multiple-parameter measures, the process is not a straightforward generalization of lower dimensional geometric problems to higher dimensional problems, because direct geometric generalization may not produce appropriate measures capable of serving the purposes of a statistical analysis. Instead, it is necessary to develop a set of meaningful measures that can constitute a well-structured system that enables further pursuit of relationships among the measures as well as the development of practical tools that facilitate interpretation and data analysis. Typical examples of such practical tools include the establishment of benchmarks for judging and interpreting measures and the search for alternative measures that reduce the computational burden. All such issues must be addressed using geometric techniques in the light of statistical considerations.

In Part I of this book, to fix conventions we recall basics of linear algebra, multi-variable calculus and Euclidean geometry in Chapters 1 and 2. In Chapter 3, we introduce the concept of normal sections, first fundamental forms and second fundamental forms. In Chapter 4, we introduce normal curvature and sectional curvatures. In Chapter 5, we study conformal transformations. This finishes our mathematical preparation.

In Part II, we first review some elementary statistics topics. In Chapters 6 and 7, we introduce basic concepts in relation to univariate distribution for discrete and continuous random variables, including the maximum likelihood estimation method. After generalizing these concepts to the bivariate and multivariate distributions in Chapter 8, we introduce simple linear regression in Chapter 9. Linear regression is the most popular statistical model, and is used as the key example in Part III of this book. To prepare for the illustration, some well-known topics in linear regression are discussed in Chapter 10.

In Part III, we apply the geometric concepts developed in Part I to the statistical issues and models articulated in Part II. The goal is to develop various measures generated by a local perturbation to assess the influence of the perturbation of model inputs. In Chapter 11, we develop the concept of likelihood displacement function. In Chapter 12, we apply the tools developed in Chapters 3 and 4 to analyze the likelihood displacement function. In the process of this development, different or mingled perspectives must be clarified. We use the regression model with some of the most well-known perspectives described in Chapter 10 for illustration. The relations among various measures that are generated in the process must also be analyzed. This is done in Chapter 13. Finally, in Chapter 14, we analyze the modification of perturbations using Chapter 5 as the theoretical foundation.

As we develop our presentation, we often encounter tedious but necessary computations. For the sake of completeness, we have placed computations in the Appendices.

Graduate students in mathematics may choose to begin reading this book starting with Part II. Likewise, graduate students in statistics may skip Part II. However, undergraduate students from both disciplines will benefit from reading this book in its entirety.

Yat-Sun Poon
Wai-Yin Poon
May, 2012

Part I

Geometry

Chapter 1

Preliminaries

1.1 Linear algebra

In this section, we recall several pre-requisite facts of linear algebra on $\mathbb{R}^n$ and the maps between them.

1.1.1 Vectors and matrices

Here we review several linear algebra facts, to set up our conventions and provide a quick reminder of elementary materials.

We denote the n-dimensional real vector space by $\mathbb{R}^n$, and treat a point in $\mathbb{R}^n$ as a column vector. A vector $\mathbf{x}$ in $\mathbb{R}^n$ is given in terms of its coordinates:

$$\mathbf{x} = \begin{pmatrix} x_1 \\ \vdots \\ x_n \end{pmatrix}, \tag{1.1}$$

where $x_i \in \mathbb{R}$ for $1 \leqslant i \leqslant n$. Given $\mathbf{x}$ and $\mathbf{y}$ in $\mathbb{R}^n$, the *vector addition* $\mathbf{x} + \mathbf{y}$ is the vector whose j-th coordinate is $x_j + y_j$ for all j from 1 to n. If λ is a real number, the *scalar multiplication* of λ on $\mathbf{x}$, denoted by $\lambda\mathbf{x}$ and $\mathbf{x}\lambda$, represents the vector whose j-th coordinate is λx_j. The vector whose coordinates are all equal to zero is called the zero-vector and is denoted by $\mathbf{0}$.

A *linear combination* of a collection of vectors $\{\mathbf{v}_1, \ldots, \mathbf{v}_k\}$ in $\mathbb{R}^n$ is a vector of the form $\lambda_1\mathbf{v}_1 + \cdots + \lambda_k\mathbf{v}_k$ for some real numbers $\lambda_1, \ldots, \lambda_k$. Note that there is not any restriction on the number k relative to the dimension n.

The *linear span* of a collection of vectors $\{\mathbf{v}_1, \ldots, \mathbf{v}_k\}$ in $\mathbb{R}^n$ is the set of all possible linear combinations for those vectors. If the linear span is equal to $\mathbb{R}^n$, we say that this collection spans $\mathbb{R}^n$.

A collection of vectors $\{\mathbf{v}_1, \ldots, \mathbf{v}_k\}$ in $\mathbb{R}^n$ is *linearly independent* if $\lambda_1\mathbf{v}_1 + \cdots + \lambda_k\mathbf{v}_k = \mathbf{0}$ implies that each λ_j is equal to zero. If this collection of vectors spans $\mathbb{R}^n$, then it is called a *basis* for $\mathbb{R}^n$. In such a case, it is necessary that $k = n$.

We use $\mathbf{e}_j$ to denote the vector whose coordinates are all equal to zero except that its j-th coordinate is equal to 1. This collection of vectors $\{\mathbf{e}_1, \ldots, \mathbf{e}_n\}$ spans $\mathbb{R}^n$ because if $\mathbf{x}$ is given as (1.1), then

$$\begin{aligned}\mathbf{x} = \begin{pmatrix} x_1 \\ x_2 \\ \vdots \\ x_n \end{pmatrix} &= x_1 \begin{pmatrix} 1 \\ 0 \\ \vdots \\ 0 \end{pmatrix} + \cdots + x_n \begin{pmatrix} 0 \\ \vdots \\ 0 \\ 1 \end{pmatrix} \\ &= x_1\mathbf{e}_1 + \cdots + x_n\mathbf{e}_n.\end{aligned}$$

It is also apparent that if $\mathbf{x} = \mathbf{0}$, then each x_j is equal to zero. Therefore, $\{\mathbf{e}_1, \ldots, \mathbf{e}_n\}$ is a basis for $\mathbb{R}^n$. We consider this particular basis as the *standard basis* for the vector space $\mathbb{R}^n$. The individual vectors are identified as *basic vectors.*

A map $\mathbf{A} : \mathbb{R}^n \to \mathbb{R}^m$ is *linear* if

$$\mathbf{A}(\lambda_1\mathbf{v}_1 + \lambda_2\mathbf{v}_2) = \lambda_1\mathbf{A}(\mathbf{v}_1) + \lambda_2\mathbf{A}(\mathbf{v}_2)$$

for all real numbers λ_1, λ_2, and vectors $\mathbf{v}_1, \mathbf{v}_2$ in $\mathbb{R}^n$. Given an $m \times n$-matrix $\mathbf{A}$, a matrix multiplication from the left on a column vector defines a linear map. To be precise,

$$\mathbf{A}(\mathbf{x}) = \begin{pmatrix} a_{11} & \ldots & a_{1n} \\ \vdots & & \vdots \\ a_{m1} & \ldots & a_{mn} \end{pmatrix} \begin{pmatrix} x_1 \\ \vdots \\ x_n \end{pmatrix}, \tag{1.2}$$

where each entry a_{ij} is a real number. In this case, we also use matrix multiplication notation $\mathbf{Ax}$ to represent the map $\mathbf{A}(x)$. This means that if

$$\begin{pmatrix} y_1 \\ \vdots \\ y_m \end{pmatrix}$$

are the coordinates of a point $\mathbf{y}$ in $\mathbb{R}^m$, then $\mathbf{Ax} = \mathbf{y}$ if and only if for all k in $\{1, \ldots, m\}$,

$$y_k = \sum_{j=1}^{n} a_{kj}x_j. \tag{1.3}$$

Because $\mathbf{Ae}_j$ is the j-th column of the matrix $\mathbf{A}$, one may re-write the above expression for the matrix $\mathbf{A}$ in terms of the result of its linear action on $\mathbf{e}_j$, i.e.,

$$\mathbf{A} = (\mathbf{Ae}_1, \ldots, \mathbf{Ae}_n)\,. \tag{1.4}$$

On the other hand, given the standard basis $\{\mathbf{e}_1, \ldots, \mathbf{e}_n\}$ and an arbitrary linear map L, by taking the coordinates of $L(\mathbf{e}_j)$ as the j-th column of a matrix the linear map L is uniquely represented by a matrix.

The *transpose* of an $m \times n$-matrix $\mathbf{A}^T$ is an $n \times m$-matrix obtained by interchanging the rows with columns in $\mathbf{A}$. To be precise, if $\mathbf{A}$ is given as in (1.2), then

$$\mathbf{A}^T = \begin{pmatrix} a_{11} & \cdots & a_{m1} \\ \vdots & & \vdots \\ a_{1n} & \cdots & a_{nm} \end{pmatrix} \tag{1.5}$$

For example, $\mathbf{x} = (x_1, \ldots, x_n)^T$ and $\mathbf{0} = (0, \ldots, 0)^T$.

An $n \times n$-matrix is also called a *square matrix*. A square matrix $\mathbf{A}$ is symmetric if $\mathbf{A} = \mathbf{A}^T$. In terms of its entries, $a_{ij} = a_{ji}$ for all $1 \leqslant i, j \leqslant n$. A square matrix $\mathbf{A}$ is *diagonal* if $a_{ij} = 0$ for all $i \neq j$. In such case, we will also denote $\mathbf{A}$ by its diagonal terms as $\mathrm{Diag}(a_{11}, \ldots, a_{nn})$. The square matrix $\mathrm{Diag}(1, \ldots, 1)$ is also denoted by $\mathbf{I}$, which is also known as the *identity matrix*.

Exercise 1.1.1 Suppose $\mathbf{A}$ is an $m \times n$-matrix and $\mathbf{B}$ is an $n \times k$-matrix, then $(\mathbf{AB})^T = \mathbf{B}^T\mathbf{A}^T$.

1.1.2 Symmetric bilinear forms

A *bilinear form* of the vector space $\mathbb{R}^n$ is a function g that associated with each pair of vectors $\mathbf{x}$ and $\mathbf{y}$, is real number $g(\mathbf{x}, \mathbf{y})$, which satisfies the following properties for all real numbers λ_1 and λ_2 and all vectors $\mathbf{x}_1, \mathbf{x}_2, \mathbf{y}_1$ and $\mathbf{y}_2$ in $\mathbb{R}^n$,

$$g(\lambda_1\mathbf{x}_1 + \lambda_2\mathbf{x}_2, \mathbf{y}) = \lambda_1 g(\mathbf{x}_1, \mathbf{y}) + \lambda_2 g(\mathbf{x}_2, \mathbf{y});$$
$$g(\mathbf{x}, \lambda_1\mathbf{y}_1 + \lambda_2\mathbf{y}_2) = \lambda_1 g(\mathbf{x}, \mathbf{y}_1) + \lambda_2 g(\mathbf{x}, \mathbf{y}_2).$$

A bilinear form is symmetric if $g(\mathbf{x}, \mathbf{y}) = g(\mathbf{y}, \mathbf{x})$ for all $\mathbf{x}$ and $\mathbf{y}$ in $\mathbb{R}^n$.

Example 1.1.2 (Dot product) For all $\mathbf{x}$ and $\mathbf{y}$ in $\mathbb{R}^n$, define

$$\langle \mathbf{x}, \mathbf{y} \rangle = \mathbf{x}^T\mathbf{y} = \sum_{j=1}^{n} x_j y_j. \tag{1.6}$$

Then $\langle \mathbf{x}, \mathbf{y} \rangle$ is the usual dot product between the vectors $\mathbf{x}$ and $\mathbf{y}$.

Example 1.1.3 (Symmetric matrices) Given an $n \times n$-matrix $\mathbf{A}$, define a map from the product $g_{\mathbf{A}} : \mathbb{R}^n \times \mathbb{R}^n$ to $\mathbb{R}$ by

$$g_{\mathbf{A}}(\mathbf{x}, \mathbf{y}) = \mathbf{x}^T\mathbf{A}\mathbf{y}. \tag{1.7}$$

The map $g_{\mathbf{A}}$ is a bilinear map associated with the matrix $\mathbf{A}$. If the matrix $\mathbf{A}$ is symmetric,

$$g_{\mathbf{A}}(\mathbf{x},\mathbf{y}) = \mathbf{x}^T\mathbf{A}\mathbf{y} = (\mathbf{x}^T\mathbf{A}\mathbf{y})^T = \mathbf{y}^T\mathbf{A}^T(\mathbf{x}^T)^T = \mathbf{y}^T\mathbf{A}\mathbf{x} = g_{\mathbf{A}}(\mathbf{y},\mathbf{x}). \quad (1.8)$$

This means that the bilinear map $g_{\mathbf{A}}$ associated with the matrix $\mathbf{A}$ is symmetric.

In particular, the dot product is the bilinear map associated with the identity matrix.

1.1.3 Vector subspaces

A subset V of $\mathbb{R}^m$ is a *vector subspace*. If for any two elements $\mathbf{v}_1$ and $\mathbf{v}_2$ in V and any two real numbers λ_1 and λ_2, $\lambda_1\mathbf{v}_1 + \lambda_2\mathbf{v}_2$ is again an element in V. Because one may choose $\lambda_1 = 0$ and $\lambda_2 = 0$, a vector subspace necessarily contains the zero vector $\mathbf{0}$ of $\mathbb{R}^m$. The *dimension* of a vector subspace V is the number of independent vectors needed to span V.

For example, the set

$$V = \{\mathbf{x} \in \mathbb{R}^{n+1} : x_{n+1} = 0\} \quad (1.9)$$

is a vector subspace. One may consider this vector subspace as a copy of $\mathbb{R}^n$. It is an n-dimensional vector subspace in $\mathbb{R}^{n+1}$.

To be precise, consider a linear map $\mathbf{A} : \mathbb{R}^n \longrightarrow \mathbb{R}^{n+1}$ defined by

$$\mathbf{A} = \begin{pmatrix} 1 & 0 & \dots & 0 & 0 \\ 0 & 1 & \dots & 0 & 0 \\ \vdots & \vdots & & \vdots & \vdots \\ 0 & 0 & \dots & 0 & 1 \\ 0 & 0 & \dots & 0 & 0 \end{pmatrix}. \quad (1.10)$$

As the image of $\mathbb{R}^n$ is equal to V, as defined above, and $\mathbf{A}\mathbf{x}_1 = \mathbf{A}\mathbf{x}_2$ if and only if $\mathbf{x}_1 = \mathbf{x}_2$, the map $\mathbf{A}$ identifies $\mathbb{R}^n$ to the vector subspace V in both one-to-one and onto fashions. We address the map $\mathbf{A}$ in (1.10) along with the subspace V the standard embedding of $\mathbb{R}^n$ in $\mathbb{R}^{n+1}$.

More generally, suppose that $\mathbf{A} : \mathbb{R}^m \longrightarrow \mathbb{R}^n$ is a linear map. Consider the image and kernel of $\mathbf{A}$.

$$\text{Image}\mathbf{A} = \{\mathbf{A}\mathbf{v} \in \mathbb{R}^n : \mathbf{v} \in \mathbb{R}^m\}, \quad \ker\mathbf{A} = \{\mathbf{v} \in \mathbb{R}^m : \mathbf{A}\mathbf{v} = 0\}.$$

This is an elementary exercise to show that Image$\mathbf{A}$ is a vector subspace of $\mathbb{R}^n$ and ker$\mathbf{A}$ is a vector subspace of $\mathbb{R}^m$.

The *rank* of an $m \times n$-matrix $\mathbf{A}$ is the dimension of vector subspace Image$\mathbf{A}$ in $\mathbb{R}^m$.

Another kind of vector spaces often encountered can be constructed as follows. Let $\mathbf{n}$ be a non-zero vector in $\mathbb{R}^{n+1}$. Define

$$V^{\perp}(\mathbf{n}) = \{\mathbf{v} \in \mathbb{R}^n : \langle \mathbf{v}, \mathbf{n} \rangle = 0\}. \tag{1.11}$$

Due to the linearity of dot product, for any real numbers λ_1 and λ_2 and vectors $\mathbf{v}_1$ and $\mathbf{v}_2$,

$$\langle \lambda_1 \mathbf{v}_1 + \lambda_2 \mathbf{v}_2, \mathbf{n} \rangle = \lambda_1 \langle \mathbf{v}_1, \mathbf{n} \rangle + \lambda_2 \langle \mathbf{v}_2, \mathbf{n} \rangle.$$

Therefore, if $\mathbf{v}_1$ and $\mathbf{v}_2$ are in the set $V^{\perp}(\mathbf{n})$, then $\lambda_1 \mathbf{v}_1 + \lambda_2 \mathbf{v}_2$ is also contained in the set $V^{\perp}(\mathbf{n})$. This shows that $V^{\perp}(\mathbf{n})$ is a vector subspace of $\mathbb{R}^n$. Because it is the set of vectors orthogonal to the vector $\mathbf{n}$, it is called the *orthogonal complement* of $\mathbf{n}$. The vector $\mathbf{n}$ is called a *normal vector* of the subspace $V^{\perp}(\mathbf{n})$.

1.1.4 Linear maps from $\mathbb{R}^n$ to $\mathbb{R}^n$

A non-zero vector $\mathbf{x}$ is an *eigenvector* of the square matrix $\mathbf{A}$ if and only if there exists a real number λ, such that $\mathbf{A}\mathbf{x} = \lambda \mathbf{x}$. The number λ is called an *eigenvalue* of the matrix $\mathbf{A}$. The set

$$V_\lambda = \{\mathbf{x} \in \mathbb{R}^n : \mathbf{A}\mathbf{x} = \lambda \mathbf{x}\}$$

is called the eigenspace of the matrix $\mathbf{A}$. All eigenspaces are *invariant subspaces* of $\mathbf{A}$, i.e., whenever $\mathbf{v}$ is in V, $\mathbf{A}\mathbf{v}$ is in V.

Not every square matrix has eigenvalus. For example, the matrix

$$J = \begin{pmatrix} 0 & -1 \\ 1 & 0 \end{pmatrix}$$

is a rotation of $90°$ on the two-dimensional plane. It does not leave any one-dimensional subspace invariant. However, we have the following.

Theorem 1.1.4 *Suppose that* $\mathbf{A} : \mathbb{R}^n \to \mathbb{R}^n$ *is a symmetric matrix, then there is a basis* $\{v_1, \ldots, v_n\}$ *for* $\mathbb{R}^n$*, such that each vector* v_j *in the basis is an eigenvector.*

A linear transformation $\mathbf{A}$ on $\mathbb{R}^n$ is *invertible* if there exists another linear transformation $\mathbf{B}$ on $\mathbb{R}^n$ such that $\mathbf{A}\mathbf{B}\mathbf{v} = \mathbf{v}$ for all $\mathbf{v} \in \mathbb{R}^n$. In terms of matrices, it is equivalent to $\mathbf{A}\mathbf{B} = \mathbf{I}$. The matrix $\mathbf{B}$ is called the *inverse* of $\mathbf{A}$. It is uniquely determined by $\mathbf{A}$ and is denoted by $\mathbf{A}^{-1}$. It can also be proved that the identity $\mathbf{A}\mathbf{A}^{-1} = \mathbf{I}$ is equivalent to $\mathbf{A}^{-1}\mathbf{A} = \mathbf{I}$.

Associated with each $n \times n$-matrix $\mathbf{A}$ are its trace and determinant. By definition, if a_{ij} is the (i, j)-th entry of the matrix $\mathbf{A}$, then the *trace* is the sum of its diagonal terms:

$$\operatorname{Tr} \mathbf{A} = \sum_{i=1}^{n} a_{ii}.$$

The *determinant* is defined inductively by n. If $\mathbf{A}$ is a one-by-one matrix, then $\det \mathbf{A}$ is equal to the value of its only entry. Assuming that the definition of the determinant of an $n \times n$-matrix is understood, we define the determinant of a $(n+1) \times (n+1)$-matrix in the following way. For each j in $\{1, \ldots, n+1\}$, let $\mathbf{A}_j$ be the $n \times n$-matrix with the first row and j-th column of the matrix $\mathbf{A}$ removed. Then

$$\det \mathbf{A} = \sum_{j=1}^{n+1} (-1)^{j+1} a_{1j} \det \mathbf{A}_j.$$

Next, we recall several important facts on trace and determinant.

Theorem 1.1.5 *Suppose that* $\mathbf{A}$ *and* $\mathbf{B}$ *are* $n \times n$*-matrices.*

- $\mathbf{A}$ *is invertible if and only if* $\det \mathbf{A} \neq 0$.
- $\det \mathbf{A} = \det \mathbf{A}^T$, $\det(\mathbf{AB}) = (\det \mathbf{A})(\det \mathbf{B})$.
- $\operatorname{Tr}(\mathbf{A} + \mathbf{B}) = \operatorname{Tr} \mathbf{A} + \operatorname{Tr} \mathbf{B}$.

If, in addition, $\mathbf{A}$ *is a symmetric matrix with a collection of eigenvectors* $\{\mathbf{v}_1, \ldots, \mathbf{v}_n\}$ *spanning* $\mathbb{R}^n$*, and with* $\{\lambda_1, \ldots, \lambda_n\}$ *as eigenvalues, then*

- $\det \mathbf{A} = \lambda_1 \cdots \lambda_n$.
- $\operatorname{Tr} \mathbf{A} = \lambda_1 + \cdots + \lambda_n$.

Example 1.1.6 Consider a 2×2-symmetric matrix

$$\mathbf{A} = \begin{pmatrix} a & b \\ b & d \end{pmatrix}.$$

By definition, $\det \mathbf{A} = ad - b^2$ and $\operatorname{Tr} \mathbf{A} = a + d$. As a symmetric matrix, the matrix $\mathbf{A}$ has two eigenvalues, λ_1 and λ_2. Assume that $\lambda_1 \geqslant \lambda_2$. Because $\det \mathbf{A} = \lambda_1 \lambda_2$ and $\operatorname{Tr} \mathbf{A} = \lambda_1 + \lambda_2$, λ_1 and λ_2 are the solution to the quadratic equation for λ_1 and λ_2 in terms of $\det \mathbf{A}$ and $\operatorname{Tr} \mathbf{A}$:

$$\lambda^2 - (\operatorname{Tr} \mathbf{A})\lambda + \det \mathbf{A} = 0.$$

It follows that

$$\begin{aligned} \lambda_1 &= \frac{1}{2}\left(\operatorname{Tr} \mathbf{A} + \sqrt{(\operatorname{Tr} \mathbf{A})^2 - 4\det \mathbf{A}}\right) \\ &= \frac{1}{2}(a + d + \sqrt{a^2 + 4b^2 + d^2 - 2ad}), \end{aligned} \tag{1.12}$$

$$\lambda_2 = \frac{1}{2}\left(\mathrm{Tr}\,\mathbf{A} - \sqrt{(\mathrm{Tr}\,\mathbf{A})^2 - 4\det\mathbf{A}}\right)$$
$$= \frac{1}{2}(a + d - \sqrt{a^2 + 4b^2 + d^2 - 2ad}). \tag{1.13}$$

Exercise 1.1.7 Suppose $\mathbf{A}$ is an invertible $n \times n$-matrix. Show that $\mathbf{A}^T$ is also invertible and $(\mathbf{A}^T)^{-1} = (\mathbf{A}^{-1})^T$.

1.1.5 A convention

In linear algebra, it is natural to consider the multiplication of an $m \times n$-matrix $\mathbf{A}$ by an $n \times k$-matrix $\mathbf{B}$ as the composition of a map from $\mathbb{R}^k$ to $\mathbb{R}^n$ with a map from $\mathbb{R}^n$ to $\mathbb{R}^m$. The result is an $m \times k$-matrix $\mathbf{AB}$.

However, we have an additional convention, i.e., if $\mathbf{v}$ is a vector in $\mathbb{R}^m$ and $\mathbf{w}$ is vector in $\mathbb{R}^n$, then we define an $m \times n$-matrix as follows:

$$\mathbf{v}\mathbf{w}^T = \begin{pmatrix} v_1 \\ \vdots \\ v_m \end{pmatrix} (w_1, \ldots, w_n) = \begin{pmatrix} v_1 w_1 & \ldots & v_1 w_n \\ \vdots & & \vdots \\ v_m w_1 & \ldots & v_m w_n \end{pmatrix}. \tag{1.14}$$

If $\mathbf{u}$ is a vector in $\mathbb{R}^n$, then

$$\mathbf{v}\mathbf{w}^T\mathbf{u} = \begin{pmatrix} v_1 \\ \vdots \\ v_m \end{pmatrix} (w_1, \ldots, w_n) \begin{pmatrix} u_1 \\ \vdots \\ u_n \end{pmatrix} = \begin{pmatrix} v_1\langle \mathbf{w}, \mathbf{u}\rangle \\ \vdots \\ v_m\langle \mathbf{w}, \mathbf{u}\rangle \end{pmatrix} = \langle \mathbf{w}, \mathbf{u}\rangle \begin{pmatrix} v_1 \\ \vdots \\ v_m \end{pmatrix}.$$

Therefore, if $\langle \mathbf{w}, \mathbf{u}\rangle = 0$, then $\mathbf{v}\mathbf{w}^T(\mathbf{u}) = 0$. If $\mathbf{u} = \lambda\mathbf{w}$ for some $\lambda \in \mathbb{R}$, then $\mathbf{v}\mathbf{w}^T\mathbf{u} = \lambda|\mathbf{w}|^2\mathbf{v}$. It is a vector in the linear span of $\mathbf{v}$. Therefore, the linear map $\mathbf{v}\mathbf{w}^T$ sends the linear span of $\mathbf{w}$ to the linear span of $\mathbf{v}$ and sends every vector in the orthogonal complement of $\mathbf{w}$ to the zero vector. In particular, $\mathbf{v}\mathbf{w}^T$ is a rank-one matrix.

1.2 Vector calculus

1.2.1 Vector-valued functions and differentials

A path $\mathbf{c}$ in $\mathbb{R}^n$ is a vector-valued function

$$\mathbf{c} : (a, b) \subseteq \mathbb{R} \to \mathbb{R}^n, \quad \mathbf{c}(t) = (x_1(t), \ldots, x_n(t))^T, \quad t \in (a, b)$$

such that each coordinate function $x_j(t)$, $j \in \{1, \ldots, n\}$, is infinitely differentiable with respect to the variable t. It passes through the point $\mathbf{x}$ if $\mathbf{c}(t_0) = \mathbf{x}$ for some t_0 in the open interval (a, b). The velocity vector of the path $\mathbf{c}(t)$ at t_0 is given by coordinate-wise differentiation with respect to t, and then an evaluation at t_0:

$$\frac{d\mathbf{c}}{dt}(t_0) = (x_1'(t_0), \ldots, x_n'(t_0))^T. \tag{1.15}$$

The collection of velocity vectors for paths through a point $\mathbf{x} \in \mathbb{R}^n$ is denoted by $T_{\mathbf{x}}\mathbb{R}^n$. This set is known as the *tangent space* of $\mathbb{R}^n$ at the point $\mathbf{x}$. Considering the n-tuples above as a vector in $\mathbb{R}^n$ that originated from the point $\mathbf{0}$ allows us to implicitly ignore the point $\mathbf{x}$ of tangency. The result is an identification of $T_{\mathbf{x}}\mathbb{R}^n$ with $\mathbb{R}^n$. We often ignore the fine distinction between $T_{\mathbf{x}}\mathbb{R}^n$ and $\mathbb{R}^n$ in computation. However, it is always helpful to know their conceptual difference.

If $\phi : \mathbb{R}^n \to \mathbb{R}^m$ is a differentiable function, it can be expressed in terms of its coordinate functions in the space $\mathbb{R}^m$, i.e.,

$$\phi(\mathbf{x}) = (\phi_1(\mathbf{x}), \ldots, \phi_m(\mathbf{x}))^T. \tag{1.16}$$

Then the differential of the map ϕ at the point $\mathbf{x}$ is the map taking the tangent space of $\mathbb{R}^n$ at $\mathbf{x}$ to the tangent space at the image point $\phi(\mathbf{x})$. This map is denoted by

$$d\phi_{\mathbf{x}} : T_{\mathbf{x}}\mathbb{R}^n \to T_{\phi(\mathbf{x})}\mathbb{R}^m. \tag{1.17}$$

If a tangent vector $\mathbf{v} = (v_1, \ldots, v_n)^T$ of $\mathbb{R}^n$ at the point $\mathbf{x}$ is given as the velocity vector of the path $\mathbf{c}(t)$, then as a matter of definition, the image vector $d\phi_{\mathbf{x}}(\mathbf{v})$ is the velocity vector of the path $\phi(\mathbf{c}(t))$ at the image point $\phi(\mathbf{x})$. To calculate this, use the Chain Rule

$$\begin{aligned}
d\phi_{\mathbf{x}}(\mathbf{v}) &= \frac{d}{dt}_{|t=0}\phi(\mathbf{c}(t)) \\
&= \left(\frac{d}{dt}_{|t=0}\phi_1(\mathbf{c}(t)), \ldots, \frac{d}{dt}_{|t=0}\phi_j(\mathbf{c}(t)), \ldots, \frac{d}{dt}_{|t=0}\phi_m(\mathbf{c}(t))\right)^T \\
&= \left(\sum_{k=1}^{n} \frac{\partial \phi_1}{\partial x_k}\frac{dx_k}{dt}, \ldots, \sum_{k=1}^{n} \frac{\partial \phi_j}{\partial x_k}\frac{dx_k}{dt}, \ldots, \sum_{k=1}^{n} \frac{\partial \phi_m}{\partial x_k}\frac{dx_k}{dt}\right)^T_{|t=0} \\
&= \left(\sum_{k=1}^{n} \frac{\partial \phi_1}{\partial x_k}v_k, \ldots, \sum_{k=1}^{n} \frac{\partial \phi_j}{\partial x_k}v_k, \ldots, \sum_{k=1}^{n} \frac{\partial \phi_m}{\partial x_k}v_k\right)^T \\
&= \begin{pmatrix} \frac{\partial \phi_1}{\partial x_1} & \frac{\partial \phi_1}{\partial x_2} & \cdots & \frac{\partial \phi_1}{\partial x_n} \\ \vdots & \vdots & & \vdots \\ \frac{\partial \phi_m}{\partial x_1} & \frac{\partial \phi_m}{\partial x_2} & \cdots & \frac{\partial \phi_m}{\partial x_n} \end{pmatrix} \begin{pmatrix} v_1 \\ v_2 \\ \vdots \\ v_n \end{pmatrix}.
\end{aligned} \tag{1.18}$$

This matrix representation reinforces the perspective that the differential of a function at the point $\mathbf{x}$ is a linear map that sends the tangent vectors of $\mathbb{R}^n$ at the point $\mathbf{x}$ to tangent vectors of $\mathbb{R}^m$ at $\phi(\mathbf{x})$.

1.2.2 Taylor expansion and extrema

Suppose that $f(\mathbf{x})$ is a real-valued function defined on an open subset U of $\mathbb{R}^n$. We treat the differential of f at $\mathbf{x}$ as a linear map

$$df_{\mathbf{x}} : T_{\mathbf{x}}\mathbb{R}^n \to T_{f(\mathbf{x})}\mathbb{R}. \tag{1.19}$$

Applying formula (1.18) to this case, we find the matrix representation of df.

$$df = \left(\frac{\partial f}{\partial x_1}, \ldots, \frac{\partial f}{\partial x_n} \right). \tag{1.20}$$

The transpose of the matrix representation for df is commonly known as the *gradient vector field.* It is denoted by ∇_f and $(df)^T$,

$$\nabla_f = \begin{pmatrix} \dfrac{\partial f}{\partial x_1} \\ \vdots \\ \dfrac{\partial f}{\partial x_n} \end{pmatrix}. \tag{1.21}$$

Note that for each $\mathbf{x}$, the column ∇_f is a vector in $T_{\mathbf{x}}\mathbb{R}^n$. The row df is a linear map from $T_{\mathbf{x}}\mathbb{R}^n$ to $T_{f(\mathbf{x})}\mathbb{R}$. In particular, if $\mathbf{v}$ is a vector in $\mathbb{R}^n$, then $df(\mathbf{v}) = \langle \nabla_f, \mathbf{v} \rangle = (\nabla_f)^T \mathbf{v}$.

The Hessian $\ddot{\mathbf{F}}$ of the function f is a matrix of second partial derivatives.

$$\ddot{\mathbf{F}} = \begin{pmatrix} \dfrac{\partial^2 f}{\partial x_1 \partial x_1} & \dfrac{\partial^2 f}{\partial x_1 \partial x_2} & \cdots & \dfrac{\partial^2 f}{\partial x_1 \partial x_n} \\ \dfrac{\partial^2 f}{\partial x_2 \partial x_1} & \dfrac{\partial^2 f}{\partial x_2 \partial x_2} & \cdots & \dfrac{\partial^2 f}{\partial x_2 \partial x_n} \\ \vdots & \vdots & & \vdots \\ \dfrac{\partial^2 f}{\partial x_n \partial x_1} & \dfrac{\partial^2 f}{\partial x_n \partial x_2} & \cdots & \dfrac{\partial^2 f}{\partial x_n \partial x_n} \end{pmatrix}. \tag{1.22}$$

It is known [?] that if the partial derivatives of f for all orders exist, then

$$\frac{\partial^2 f}{\partial x_i \partial x_j} = \frac{\partial^2 f}{\partial x_j \partial x_i}$$

for all $1 \leqslant i, j \leqslant n$. It follows that $\ddot{\mathbf{F}}$ is a symmetric matrix.

Theorem 1.2.1 (Taylor expansion) *Suppose that f is a function defined on an open set $U \subseteq \mathbb{R}^n$. If all its partial derivatives with any order exist, then for any $\mathbf{x}_0$ in U and $\mathbf{v}$ in $\mathbb{R}^n$,*

$$f(\mathbf{x}+\mathbf{v}) = f(\mathbf{x}) + df(\mathbf{v}) + \frac{1}{2}\mathbf{v}^T\ddot{\mathbf{F}}(\mathbf{x}_0)\mathbf{v} + R(\mathbf{x}_0, \mathbf{v}) \tag{1.23}$$

where

$$\lim_{|\mathbf{v}|\to 0} \frac{R(\mathbf{x}_0, \mathbf{v})}{|\mathbf{v}|^2} = 0.$$

In all our subsequent discussion, we assume the partial derivatives of our functions exist to all orders.

1.2.3 Extrema and Lagrange multiplier theorem

If a real-valued function $f(\mathbf{x})$ defined on an open subset U of $\mathbb{R}^n$ attains it local maximum or local minimum at the point $\mathbf{x}_0 \in U$, then $df_{\mathbf{x}_0} = \mathbf{0}$.

A point $\mathbf{x}_0 \in U$ is a *critical point* of $f(\mathbf{x})$ if $df_{\mathbf{x}_0} = \mathbf{0}$.

To determine when a critical point $\mathbf{x}_0$ is a local maximum and local minimum, we appeal to the Taylor expansion of $f(\mathbf{x})$ for the point $\mathbf{x}_0$. Assuming $\mathbf{x}_0$ is a critical point, according to Theorem 1.2.1 when $|\mathbf{v}|$ is small, one can estimate

$$f(\mathbf{x}_0+\mathbf{v}) \simeq f(\mathbf{x}_0) + 0 + \frac{1}{2}\mathbf{v}^T\ddot{\mathbf{F}}(\mathbf{x}_0)\mathbf{v}. \tag{1.24}$$

Therefore, f attains a local minimum at $\mathbf{x}_0$ if and only if $\mathbf{v}^T\ddot{\mathbf{F}}(\mathbf{x}_0)\mathbf{v} \geqslant 0$ for any $\mathbf{v}$. Because $\ddot{\mathbf{F}}(\mathbf{x}_0)$ is a symmetric matrix, this condition is equivalent to requiring all eigenvalues of $\ddot{\mathbf{F}}$ at $\mathbf{x}_0$ to be non-negative. In the event that $f(\mathbf{x}_0)$ is strictly bigger than $f(\mathbf{x}_0+\mathbf{v})$ for a sufficiently small $\mathbf{v}$, it becomes necessary and sufficient for all eigenvalues of $\ddot{\mathbf{F}}$ at $\mathbf{x}_0$ to be strictly positive. In other words, the Hessian $\ddot{\mathbf{F}}(\mathbf{x}_0)$ is *positive-definite*.

An identical argument shows that a local maximum for $f(\mathbf{x})$ occurs at $\mathbf{x}_0$ when $\ddot{\mathbf{F}}(\mathbf{x}_0)$ is negative-definite.

In future, we shall seek extrema or critical points for a real-valued function $f(\mathbf{x})$, not within an open set in $\mathbb{R}^n$, but among the points constrained by a different real-valued function. Let $g(\mathbf{x})$ be a differentiable real-valued function defined on the open set U. Let $S = \{\mathbf{x} \in U : g(\mathbf{x}) = c\}$ where c is a constant. To seek the maximum of $f(\mathbf{x})$ restricted to the set S, we will need the following concepts.

Theorem 1.2.2 (Lagrange Multiplier) *Suppose that $f(\mathbf{x})$ and $g(\mathbf{x})$ are real-valued functions defined on an open subset U of $\mathbb{R}^n$. Let $S = \{\mathbf{x} \in U : g(\mathbf{x}) =$*

$c\}$. *Assuming that* $\mathbf{x}_0 \in S$ *and* $\nabla_g(\mathbf{x}_0) \neq 0$, *then the point* $\mathbf{x}_0$ *is a critical point of* $f(\mathbf{x})$ *subjected to the constraint* S *if there is a constant* λ, *such that*

$$\nabla_f(\mathbf{x}_0) = \lambda \nabla_g(\mathbf{x}_0). \tag{1.25}$$

The constant λ *is called a Lagrange multiplier.*

As an exercise of the Lagrange Multiplier Theorem, we prove the following propositions.

Proposition 1.2.3 *Let* S *be the set of unit vectors in* $\mathbb{R}^n$. *Let* $\mathbf{A}$ *be a symmetric* $n \times n$*-matrix. Define a function* $f : S \to \mathbb{R}$ *by* $f(\mathbf{x}) = \mathbf{x}^T\mathbf{A}\mathbf{x}$. *A point* $\mathbf{x}_0$ *is a critical point of the function* ϕ *on* S *if and only if it is an eigenvector of the matrix* $\mathbf{A}$. *The value of the function* $f(\mathbf{x})$ *at the critical point* $\mathbf{x}_0$ *is the corresponding eigenvalue.*

Proof: Let $g : \mathbb{R}^n \to \mathbb{R}$ be defined by $g(\mathbf{x}) = \langle \mathbf{x}, \mathbf{x} \rangle$, then $S = \{\mathbf{x} \in \mathbb{R}^n : g(\mathbf{x}) = 1\}$. In terms of coordinates on $\mathbb{R}^n$, $\frac{\partial}{\partial x_j}\mathbf{x} = \mathbf{e}_j$. Then

$$\frac{\partial}{\partial x_j} g(\mathbf{x}) = \left(\frac{\partial}{\partial x_j}\mathbf{x}\right)^T \mathbf{x} + \mathbf{x}^T\left(\frac{\partial}{\partial x_j}\mathbf{x}\right) = \mathbf{e}_j^T\mathbf{x} + \mathbf{x}^T\mathbf{e}_j = 2\mathbf{e}_j^T\mathbf{x} = 2x_j. \tag{1.26}$$

Therefore, $\nabla_g = 2\mathbf{x}$. In particular $\nabla_g \neq \mathbf{0}$ when $\mathbf{x}$ is in the set S.

Consider $f(\mathbf{x}) = \mathbf{x}^T\mathbf{A}\mathbf{x}$ as a function defined on $\mathbb{R}^n$. Then

$$\frac{\partial}{\partial x_j} f(\mathbf{x}) = \left(\frac{\partial}{\partial x_j}\mathbf{x}\right)^T \mathbf{A}\mathbf{x} + \mathbf{x}^T\mathbf{A}\left(\frac{\partial}{\partial x_j}\mathbf{x}\right) = \mathbf{e}_j^T\mathbf{A}\mathbf{x} + \mathbf{x}^T\mathbf{A}\mathbf{e}_j. \tag{1.27}$$

Because $\mathbf{A}$ is a symmetric matrix,

$$\frac{\partial}{\partial x_j} f(\mathbf{x}) = 2\mathbf{e}_j^T\mathbf{A}\mathbf{x}. \tag{1.28}$$

According to Theorem 1.2.2, a point $\mathbf{x}_0$ on S is a critical point of f subjected to the constraint S if and only if there exists a Lagrange multiplier λ, such that at $\mathbf{x}_0$ $\nabla_f = \lambda\nabla_g$. As a consequence of Equations (1.26) and (1.28),

$$\langle \mathbf{e}_j, \mathbf{A}\mathbf{x}_0 \rangle = \mathbf{e}_j^T\mathbf{A}\mathbf{x}_0 = \lambda\mathbf{e}_j^T\mathbf{x}_0 = \langle \mathbf{e}_j, \lambda\mathbf{x}_0 \rangle$$

for each $j = 1, \ldots, n$. This means $\langle \mathbf{e}_j, \mathbf{A}\mathbf{x}_0 - \lambda\mathbf{x}_0 \rangle = 0$ for all basic vectors. It is possible only if $\mathbf{A}\mathbf{x}_0 - \lambda\mathbf{x}_0 = \mathbf{0}$.

Furthermore, because $\langle \mathbf{x}_0, \mathbf{x}_0 \rangle = 1$, $f(\mathbf{x}_0) = \mathbf{x}_0^T\mathbf{A}\mathbf{x} = \lambda\mathbf{x}_0^T\mathbf{x}_0 = \lambda$. ■

Proposition 1.2.4 *Let* S *be the set of unit vectors in* $\mathbb{R}^n$. *Let* $\mathbf{A}$ *be a symmetric* $n \times n$*-matrix. Define a function* $f : S \to \mathbb{R}$ *by* $f(\mathbf{x}) = \mathbf{x}^T\mathbf{A}\mathbf{x}$. *There exists eigenvectors* $\mathbf{v}_1, \ldots, \mathbf{v}_n$ *with eigenvalues* λ_j, *such that* $|\mathbf{v}_j|^2 = 1$, $\langle \mathbf{v}_j, \mathbf{v}_k \rangle = 0$ *if* $j \neq k$ *and* $\lambda_{j+1} \geqslant \lambda_j$ *for all* $1 \leqslant j \leqslant n-1$.

Proof: For the function $f(\mathbf{x})$, let $\mathbf{v}_1$ be the point at which $f(\mathbf{x})$ attains its global maximum on S. By the last proposition, $\mathbf{v}_1$ is an eigenvector and the maximum value $f(\mathbf{v}_1) = \lambda_{\max} = \lambda_1$ is the corresponding eigenvalue. Let $(\mathbb{R}^n)^{\perp}(\mathbf{v}_1)$ be the orthogonal complement of $\mathbf{v}_1$ in $\mathbb{R}^n$. Its intersection with S is a sphere in one dimension less than that of S. To find the global maximum of f on $S \cap (\mathbb{R}^n)^{\perp}(\mathbf{v}_1)$, we get $\mathbf{v}_2$, which is another eigenvector of the matrix $\mathbf{A}$ with eigenvalue λ_2. By construction, $\lambda_1 \geqslant \lambda_2$ and $\mathbf{v}_2$ is orthogonal to $\mathbf{v}_1$. Inductively, when we get a collection of mutually orthogonal eigenvectors $\{\mathbf{v}_1, \ldots, \mathbf{v}_k\}$, let $(\mathbb{R}^n)^{\perp}(\mathbf{v}_1, \ldots, \mathbf{v}_k)$ be the orthogonal complement of the linear span of $\{\mathbf{v}_1, \ldots, \mathbf{v}_k\}$ and one finds $\mathbf{v}_{k+1}$ as the global maximum of $f(\mathbf{x})$ on $S \cap (\mathbb{R}^n)^{\perp}(\mathbf{v}_1, \ldots, \mathbf{v}_k)$. This inductive process is carried out until we exhaust the space S. ■

Chapter 2

Euclidean Geometry

In this chapter, we examine several kinds of mappings from $\mathbb{R}^n$ to itself. These are studied with regards to the geometry dictated by the inner product defined on $\mathbb{R}^n$.

We use $\langle \mathbf{x}, \mathbf{y} \rangle$ to denote the dot product between two vectors $\mathbf{x}, \mathbf{y}$ in $\mathbb{R}^n$. Then

- The length of a vector $\mathbf{x}$ in $\mathbb{R}^n$ is $|\mathbf{x}| = \sqrt{\langle \mathbf{x}, \mathbf{x} \rangle}$.
- The distance between two vectors $\mathbf{x}$ and $\mathbf{y}$ in $\mathbb{R}^n$ is $|\mathbf{x} - \mathbf{y}|$.
- If $\mathbf{x}$ and $\mathbf{y}$ are two non-zero vectors in $\mathbb{R}^n$, then the angle between them is the number θ in $[0, 2\pi)$, such that $\cos\theta = \frac{\langle \mathbf{x}, \mathbf{y} \rangle}{|\mathbf{x}||\mathbf{y}|}$.
- A pair of vectors $\mathbf{x}$ and $\mathbf{y}$ are said to be orthogonal or perpendicular if $\langle \mathbf{x}, \mathbf{y} \rangle = 0$.

2.1 Orthogonal transformations

A map F from $\mathbb{R}^n$ to $\mathbb{R}^n$ is an **isometry** if, for all vectors $\mathbf{x}$ and $\mathbf{y}$ in $\mathbb{R}^n$,

$$|F(\mathbf{x}) - F(\mathbf{y})| = |\mathbf{x} - \mathbf{y}|. \tag{2.1}$$

We see that while isometries from $\mathbb{R}^n$ to $\mathbb{R}^n$ are not necessarily linear maps, some linear maps are isometries.

Proposition 2.1.1 *A linear map* $\mathbf{A}$ *on* $\mathbb{R}^n$ *to itself is an isometry if and only if, for all vectors* $\mathbf{x}$, $\mathbf{y}$ *in* $\mathbb{R}^n$, $\langle \mathbf{A}\mathbf{x}, \mathbf{A}\mathbf{y} \rangle = \langle \mathbf{x}, \mathbf{y} \rangle$.

Proof: Suppose that a linear map satisfies $\langle \mathbf{A}\mathbf{x}, \mathbf{A}\mathbf{y} \rangle = \langle \mathbf{x}, \mathbf{y} \rangle$, then

$|\mathbf{A}\mathbf{x} - \mathbf{A}\mathbf{y}|^2 = |\mathbf{A}(\mathbf{x} - \mathbf{y})|^2 = \langle \mathbf{A}(\mathbf{x} - \mathbf{y}), \mathbf{A}(\mathbf{x} - \mathbf{y}) \rangle = \langle \mathbf{x} - \mathbf{y}, \mathbf{x} - \mathbf{y} \rangle = |\mathbf{x} - \mathbf{y}|^2$.

It follows that $\mathbf{A}$ is an isometry.

Suppose that $\mathbf{A}$ is an isometry, then $|\mathbf{A}\mathbf{x} - \mathbf{A}\mathbf{y}|^2 = |\mathbf{x} - \mathbf{y}|^2$ for any pair of vectors $\mathbf{x}$ and $\mathbf{y}$ in $\mathbb{R}^n$. This identity can be expanded to

$$|\mathbf{A}\mathbf{x}|^2 - 2\langle \mathbf{A}\mathbf{x}, \mathbf{A}\mathbf{y} \rangle + |\mathbf{A}\mathbf{y}|^2 = |\mathbf{x}|^2 - 2\langle \mathbf{x}, \mathbf{y} \rangle + |\mathbf{y}|^2.$$

Because $\mathbf{A}$ is also linear, $\mathbf{A}(\mathbf{0}) = \mathbf{0}$. Applying the above formula to the pair of vectors $\mathbf{x}$ and $\mathbf{0}$, we find that $|\mathbf{Ax}|^2 = |\mathbf{x}|^2$ for all vectors $\mathbf{x}$ in $\mathbb{R}^n$. In particular, we also have $|\mathbf{Ay}|^2 = |\mathbf{y}|^2$. The identity above then implies that $\langle \mathbf{Ax}, \mathbf{Ay} \rangle = \langle \mathbf{x}, \mathbf{y} \rangle$, as claimed. ■

As noted in the proof of the above proposition, $|\mathbf{Ax}| = |\mathbf{x}|$ if $\mathbf{A}$ is a linear isometry. It follows that an isometry preserves the length of vectors and the distance between vectors. In addition,

$$\frac{\langle \mathbf{Ax}, \mathbf{Ay} \rangle}{|\mathbf{Ax}||\mathbf{Ay}|} = \frac{\langle \mathbf{x}, \mathbf{y} \rangle}{|\mathbf{x}||\mathbf{y}|}.$$

A linear isometry also preserves angles between every pair of non-zero vectors.

When a linear isometry is given by a matrix $\mathbf{A}$, then $\mathbf{x}^T\mathbf{A}^T\mathbf{Ay} = \mathbf{x}^T\mathbf{y}$ for all vectors $\mathbf{x}$ and $\mathbf{y}$. This is possible if and only if $\mathbf{A}^T\mathbf{A} = \mathbf{I}$.

The *orthogonal group* $O(n)$ of $\mathbb{R}^n$ is a subset of $n \times n$-matrices

$$O(n) = \{\mathbf{A} : \mathbf{A}^T\mathbf{A} = \mathbf{I}\}. \tag{2.2}$$

Elements in this set are called orthogonal matrices, and the Linear maps represented by such matrices are called orthogonal transformations.

Note that if $\mathbf{A}$ and $\mathbf{B}$ are in $O(n)$, so is $\mathbf{AB}$ because

$$(\mathbf{AB})^T\mathbf{AB} = \mathbf{B}^T\mathbf{A}^T\mathbf{AB} = \mathbf{B}^T\mathbf{IB} = \mathbf{B}^T\mathbf{B} = \mathbf{I}.$$

By definition if $\mathbf{A}$ is in $O(n)$, then it has an inverse, namely $\mathbf{A}^{-1} = \mathbf{A}^T$. In addition, $\mathbf{A}^{-1}$ is also in $O(n)$. Combining the above shows that the set $O(n)$ forms a *group*.

Exercise 2.1.2 If F_1 and F_2 are both isometries from $\mathbb{R}^n$ to $\mathbb{R}^n$, so is their composition $F_2 \circ F_1$.

Exercise 2.1.3 (i) The column vectors of orthogonal matrices are mutually orthogonal. (ii) The length of each column vector is equal to one.

Exercise 2.1.4 If $\mathbf{A}$ is an orthogonal matrix, then $(\det \mathbf{A})^2 = 1$.

2.2 Rigid motions

Choose a vector $\mathbf{p}$ in $\mathbb{R}^n$. The *translation* by $\mathbf{p}$ is the map $F_\mathbf{p}$ from $\mathbb{R}^n$ to $\mathbb{R}^n$, defined by

$$F_\mathbf{p}(\mathbf{v}) = \mathbf{p} + \mathbf{v} \tag{2.3}$$

for all $\mathbf{v}$ in $\mathbb{R}^n$. Because

$$F_\mathbf{p}(\mathbf{x}) - F_\mathbf{p}(\mathbf{y}) = \mathbf{x} - \mathbf{y}$$

for all $\mathbf{x}$ and $\mathbf{y}$ in $\mathbb{R}^n$, translations are isometries. However, when $F_{\mathbf{p}}(\mathbf{0}) = \mathbf{p}$, translation by a non-zero vector field is not a linear map. In particular, it can not be represented by a matrix, let alone an orthogonal matrix.

Suppose that $\mathbf{A}$ is an orthogonal matrix and $F_{\mathbf{p}}$ is a translation by a vector $\mathbf{p}$ in $\mathbb{R}^n$. We can consider their compositions $\mathbf{A} \circ F_{\mathbf{p}}$ and $F_{\mathbf{p}} \circ \mathbf{A}$. For any vector $\mathbf{x}$ in $\mathbb{R}^n$,

$$\mathbf{A} \circ F_{\mathbf{p}}(\mathbf{x}) = \mathbf{A}(F_{\mathbf{p}}(\mathbf{x})) = \mathbf{A}(\mathbf{p} + \mathbf{x}) = \mathbf{A}\mathbf{p} + \mathbf{A}\mathbf{x}.$$
$$F_{\mathbf{p}} \circ \mathbf{A}(\mathbf{x}) = F_{\mathbf{p}}(\mathbf{A}\mathbf{x}) = \mathbf{p} + \mathbf{A}\mathbf{x}.$$

It follows that $\mathbf{A} \circ F_{\mathbf{p}} \neq F_{\mathbf{p}} \circ \mathbf{A}$. Instead,

$$\mathbf{A} \circ F_{\mathbf{p}} = F_{\mathbf{A}\mathbf{p}} \circ \mathbf{A}. \tag{2.4}$$

Equivalently,

$$\mathbf{A} \circ F_{\mathbf{p}} \circ \mathbf{A}^{-1} = F_{\mathbf{A}\mathbf{p}}.$$

We have seen that a translation fails to be a linear map because it sends the zero vector to a non-zero vector. The next lemma shows that this is the only way an isometry on $\mathbb{R}^n$ fails to be an orthogonal transformation.

Lemma 2.2.1 *If F is an isometry on $\mathbb{R}^n$, such that $F(\mathbf{0}) = \mathbf{0}$, then it is an orthogonal transformation.*

Proof: Because F is an isometry,

$$|F(\mathbf{x}) - F(\mathbf{y})| = |\mathbf{x} - \mathbf{y}| \tag{2.5}$$

for all $\mathbf{x}$, $\mathbf{y}$ in $\mathbb{R}^n$. Because $F(\mathbf{0}) = \mathbf{0}$, when the above identity is applied to the case in which $\mathbf{y} = \mathbf{0}$, we have

$$|F(\mathbf{x})| = |\mathbf{x}|$$

for all $\mathbf{x}$ in $\mathbb{R}^n$. Apply the last observation to the left-hand side of the identity (2.5) and we have

$$\begin{aligned}|F(\mathbf{x}) - F(\mathbf{y})|^2 &= \langle F(\mathbf{x}) - F(\mathbf{y}), F(\mathbf{x}) - F(\mathbf{y})\rangle \\ &= \langle F(\mathbf{x}), F(\mathbf{x})\rangle + \langle F(\mathbf{y}), F(\mathbf{y})\rangle - 2\langle F(\mathbf{x}), F(\mathbf{y})\rangle \\ &= |\mathbf{x}|^2 + |\mathbf{y}|^2 - 2\langle F(\mathbf{x}), F(\mathbf{y})\rangle.\end{aligned}$$

According to (2.5), it is equal to

$$|\mathbf{x} - \mathbf{y}|^2 = \langle \mathbf{x} - \mathbf{y}, \mathbf{x} - \mathbf{y}\rangle = |\mathbf{x}|^2 - 2\langle \mathbf{x}, \mathbf{y}\rangle + |\mathbf{y}|^2.$$

It follows that F preserves the inner product:

$$\langle F(\mathbf{x}), F(\mathbf{y})\rangle = \langle \mathbf{x}, \mathbf{y}\rangle. \tag{2.6}$$

Consider the image of the standard basis $\{F\mathbf{e}_1, \ldots, F\mathbf{e}_n\}$. Because F preserves the inner product, $\{F\mathbf{e}_1, \ldots, F\mathbf{e}_n\}$ is an orthonormal basis. According to (2.6) above, for all real numbers λ_1, λ_2, and all vectors $\mathbf{x}_1$, $\mathbf{x}_2$,

$$\begin{aligned}\langle F(\lambda_1\mathbf{x}_1 + \lambda_2\mathbf{x}_2), F(\mathbf{e}_j)\rangle &= \langle \lambda_1\mathbf{x}_1 + \lambda_2\mathbf{x}_2, \mathbf{e}_j\rangle = \lambda_1\langle \mathbf{x}_1, \mathbf{e}_j\rangle + \lambda_2\langle \mathbf{x}_2, \mathbf{e}_j\rangle,\\ \langle \lambda_1 F(\mathbf{x}_1) + \lambda_2 F(\mathbf{x}_2), F(\mathbf{e}_j)\rangle &= \lambda_1\langle F(\mathbf{x}_1), F(\mathbf{e}_j)\rangle + \lambda_2\langle F(\mathbf{x}_2), F(\mathbf{e}_j)\rangle\\ &= \lambda_1\langle \mathbf{x}_1, \mathbf{e}_j\rangle + \lambda_2\langle \mathbf{x}_2, \mathbf{e}_j\rangle.\end{aligned}$$

It follows that

$$\langle F(\lambda_1\mathbf{x}_1 + \lambda_2\mathbf{x}_2), F(\mathbf{e}_j)\rangle = \langle \lambda_1 F(\mathbf{x}_1) + \lambda_2 F(\mathbf{x}_2), F(\mathbf{e}_j)\rangle$$

for each j. Because $\{F\mathbf{e}_1, \ldots, F\mathbf{e}_n\}$ is a basis, it is possible if and only if

$$F(\lambda_1\mathbf{x}_1 + \lambda_2\mathbf{x}_2) = \lambda_1 F(\mathbf{x}_1) + \lambda_2 F(\mathbf{x}_2).$$

Therefore, F is a linear map. Because it also preserves the inner product, it is also an orthogonal transformation. ■

Theorem 2.2.2 *For every isometry F on $\mathbb{R}^n$, there exists a unique orthogonal transformation $\mathbf{A}$ and a vector $\mathbf{p}$, such that $F = F_{\mathbf{p}} \circ \mathbf{A}$.*

Proof: Let $\mathbf{p} = F(\mathbf{0})$. Define an isometry $\mathbf{A}$ by a composition $\mathbf{A} = F_{-\mathbf{p}} \circ F$. It follows that

$$\mathbf{A}(\mathbf{0}) = F_{-\mathbf{p}}(F(\mathbf{0})) = -\mathbf{p} + F(\mathbf{0}) = -\mathbf{p} + \mathbf{p} = \mathbf{0}.$$

The last lemma shows that $\mathbf{A}$ is an orthogonal transformation. Because $F_{\mathbf{p}} \circ F_{-\mathbf{p}} = \mathbf{I}$,

$$F_{\mathbf{p}} \circ \mathbf{A} = F_{\mathbf{p}} \circ F_{-\mathbf{p}} \circ F = F$$

as claimed. ■

2.3 Translation of vector subspaces

Because a vector subspace must contain the origin, one may consider the vector subspace in $\mathbb{R}^n$ as a plane with a certain dimension going through the origin. Therefore, there are many subspaces of $\mathbb{R}^n$ that fail to be vector subspaces. For instance, curved objects such as spheres are not vector subspaces. The planes in $\mathbb{R}^n$ that do not passing through the origin are also not vector subspaces neither.

The translation of the set S in $\mathbb{R}^n$ by a non-zero vector $\mathbf{p}$ is the set

$$F_{\mathbf{p}}(S) = \{\mathbf{p} + \mathbf{v} : \mathbf{v} \in S\}. \tag{2.7}$$

For example, take a vector subspace $V^{\perp}(\mathbf{n})$ with normal vector $\mathbf{n}$. Let $\mathbf{p}$ be a vector not contained in $V^{\perp}(\mathbf{n})$. The translation of $V^{\perp}(\mathbf{n})$ by the vector $\mathbf{p}$ is the set

$$W = \{\mathbf{p} + \mathbf{v} : \langle \mathbf{v}, \mathbf{n} \rangle = 0\}. \tag{2.8}$$

This set fails to be a vector subspace because it does not contain the origin.

In general, whenever V is a vector subspace in $\mathbb{R}^n$ and $\mathbf{p}$ is not contained in V, then $W = F_{\mathbf{p}}(V)$ is not a vector subspace. However, we may remain interested in studying the geometry in $F_{\mathbf{p}}(V)$ by taking advantage of the fact that it can be translated back to a vector subspace using the vector $-\mathbf{p}$.

For instance, if $\mathbf{w}_1$ and $\mathbf{w}_2$ are in W, then there exists $\mathbf{v}_1$ and $\mathbf{v}_2$ such that

$$\mathbf{w}_1 = \mathbf{p} + \mathbf{v}_1, \quad \mathbf{w}_2 = \mathbf{p} + \mathbf{v}_2.$$

We consider the distance and angle between the two points $\mathbf{w}_1$ and $\mathbf{w}_2$ in terms of the two vectors $\mathbf{v}_1 = F_{-\mathbf{p}}\mathbf{w}_1$, $\mathbf{v}_2 = F_{-\mathbf{p}}\mathbf{w}_2$, which originated from the zero vector $\mathbf{0}$ in $\mathbb{R}^n$.

For instance, the distance d and angle θ between $\mathbf{w}_1$ and $\mathbf{w}_2$ "within" the space W are respectively defined as

$$d(\mathbf{w}_1, \mathbf{w}_2) = |F_{-\mathbf{p}}\mathbf{w}_1 - F_{-\mathbf{p}}\mathbf{w}_2|, \quad \theta(\mathbf{w}_1, \mathbf{w}_2) = \theta(F_{-\mathbf{p}}\mathbf{w}_1, F_{-\mathbf{p}}\mathbf{w}_2). \tag{2.9}$$

Example 2.3.1 In $\mathbb{R}^3$ with coordinates x_1, x_2, x_3. A vertical cylinder is given by the equation

$$x_1^2 + x_2^2 = 1.$$

The point $\mathbf{p} = (1, 0, 0)^T$ is on the cylinder. The functions

$$\alpha(t) = (\cos t, \sin t, t)^T, \quad \beta(t) = (1, 0, t)^T$$

describe two different paths on the cylinder that both pass through the point $\mathbf{p}$ when $t = 0$. The velocity vectors at $t = 0$ are

$$\mathbf{v}_1 = \frac{d\alpha}{dt}_{|t=0} = \begin{pmatrix} -\sin t \\ \cos t \\ 1 \end{pmatrix}_{|t=0} = \begin{pmatrix} 0 \\ 1 \\ 1 \end{pmatrix}, \quad \mathbf{v}_2 = \frac{d\beta}{dt}_{|t=0} = \begin{pmatrix} 0 \\ 0 \\ 1 \end{pmatrix}.$$

In the above expressions, we consider $\mathbf{v}_1$ and $\mathbf{v}_2$ as vectors in $\mathbb{R}^3$. However, if one sketches these objects with velocity vectors at the point $\mathbf{p}$ originating from $\mathbf{p}$, one would sketch $F_{\mathbf{p}}(\mathbf{v}_1)$ and $F_{\mathbf{p}}(\mathbf{v}_2)$. On the other hand, when one considers the angle between these vectors as tangent to the cylinder at the point $\mathbf{p}$, one actually calculates the angles between $\mathbf{v}_1$ and $\mathbf{v}_2$.

2.4 Conformal transformations

A linear transformation $\mathbf{L}$ is a *conformal map* if and only if there exists a constant $c > 0$, such that for any vectors $\mathbf{x}$, $\mathbf{y}$ in $\mathbb{R}^n$,

$$\langle \mathbf{Lx}, \mathbf{Ly} \rangle = c^2 \langle \mathbf{x}, \mathbf{y} \rangle. \tag{2.10}$$

In terms of matrices, it is equivalent to require

$$\mathbf{x}^T \mathbf{L}^T \mathbf{Ly} = c^2 \mathbf{x}^T \mathbf{y}$$

for all $\mathbf{x}$ and $\mathbf{y}$. This is possible if and only if $\mathbf{L}^T\mathbf{L} = c^2\mathbf{I}$. Such matrices are called *conformal matrices.*

Because $|\mathbf{Lx}|^2 = \langle \mathbf{Lx}, \mathbf{Lx} \rangle = c^2 |\mathbf{x}|^2$, when $c \neq 1$, $\mathbf{L}$ fails to be an isometry and does not preserve distance. However, $|\mathbf{Lx}| = c|\mathbf{x}|$ for all $\mathbf{x}$ and equation (2.10) implies that for all $\mathbf{x}$ and $\mathbf{y}$,

$$\frac{\langle \mathbf{Lx}, \mathbf{Ly} \rangle}{|\mathbf{Lx}||\mathbf{Ly}|} = \frac{c^2 \langle \mathbf{x}, \mathbf{y} \rangle}{c|\mathbf{x}|c|\mathbf{y}|} = \frac{\langle \mathbf{x}, \mathbf{y} \rangle}{|\mathbf{x}||\mathbf{y}|}.$$

Therefore, conformal maps preserve angles between all non-zero vectors.

The collection of conformal maps is denoted by

$$CO(n) = \{\mathbf{L} : \mathbf{L}^T\mathbf{L} = c^2\mathbf{I} \text{ for some } c > 0\}. \tag{2.11}$$

If $\mathbf{A}^T\mathbf{A} = a^2\mathbf{I}$ and $\mathbf{B}^T\mathbf{B} = b^2\mathbf{I}$ for some $a > 0$ and $b > 0$, then $(\mathbf{AB})^T(\mathbf{AB}) = (ab)^2\mathbf{I}$. Therefore, if $\mathbf{A}, \mathbf{B}$ are in $CO(n)$, The same is true for $\mathbf{AB}$.

Exercise 2.4.1 Show that if $\mathbf{A}$ is in $CO(n)$, then $\mathbf{A}$ is an invertible matrix and its inverse $\mathbf{A}^{-1}$ is also in $CO(n)$.

2.5 Orthonormal basis

A set of non-zero vectors $\{\mathbf{w}_1, \ldots, \mathbf{w}_r\}$ in a vector subspace W of $\mathbb{R}^n$ is an *orthogonal basis* if it is a basis, and if the vectors are mutually orthogonal, i.e., $\langle \mathbf{w}_i, \mathbf{w}_j \rangle = 0$ for all $i \neq j$ and $1 \leqslant i, j \leqslant r$. If in addition, the length of each $\mathbf{w}_j$ is equal to one, then the basis is said to be an *orthonormal basis.*

For example, the standard basis $\{\mathbf{e}_1, \ldots, \mathbf{e}_n\}$ is an orthonormal basis for $\mathbb{R}^n$. Note that if $\{\mathbf{w}_1, \ldots, \mathbf{w}_r\}$ is an orthogonal basis for a vector subspace W, then one can *normalize* the length of each vector to get an orthonormal basis. To be precise, the collection of vectors

$$\frac{1}{|\mathbf{w}_1|}\mathbf{w}_1, \ldots, \frac{1}{|\mathbf{w}_r|}\mathbf{w}_r$$

is an orthonormal basis.

In addition, if the collection of non-zero vectors $\{\mathbf{w}_1, \ldots, \mathbf{w}_r\}$ is mutually orthogonal, then they are linearly independent because if there are constants λ_j, $1 \leqslant j \leqslant r$ such that

$$\lambda_1 \mathbf{w}_1 + \cdots + \lambda_r \mathbf{w}_r = \mathbf{0},$$

then for each j,

$$0 = \langle \mathbf{w}_j, \mathbf{0} \rangle = \langle \mathbf{w}_j, \lambda_1 \mathbf{w}_1 + \cdots + \lambda_r \mathbf{w}_r \rangle = \lambda_j \langle \mathbf{w}_j, \mathbf{w}_j \rangle.$$

This is only possible only when $\lambda_j = 0$ for each j. In particular, the dimension of the linear span of $\{\mathbf{w}_1, \ldots, \mathbf{w}_r\}$ is equal to r.

Lemma 2.5.1 (Gram-Schmidt Process) *For each vector subspace W, there exists an orthonormal basis.*

Proof: Let $\mathbf{v}_1$ be a non-zero vector in W. If $\dim W = 1$, then $\mathbf{w}_1 = \frac{1}{|\mathbf{v}_1|}\mathbf{v}_1$ forms an orthonormal basis for W. If $\mathbf{w}_1$ does not span W, then there exists $\mathbf{v} \in W$ such that $\mathbf{w}_1$ and $\mathbf{v}$ are linearly independent. Let

$$\mathbf{v}_2 = \mathbf{v} - \langle \mathbf{v}, \mathbf{w}_1 \rangle \mathbf{w}_1.$$

Because $\mathbf{v}$ and $\mathbf{w}_1$ are in W, the same is true for $\mathbf{v}_2$. Because $\mathbf{v}$ and $\mathbf{w}_1$ are linearly independent, so is the collection of vectors $\{\mathbf{w}_1, \mathbf{v}_2\}$. In addition,

$$\langle \mathbf{v}_2, \mathbf{w}_1 \rangle = \langle \mathbf{v}, \mathbf{w}_1 \rangle - \langle \mathbf{v}, \mathbf{w}_1 \rangle \langle \mathbf{w}_1, \mathbf{w}_1 \rangle.$$

Because $\langle \mathbf{w}_1, \mathbf{w}_1 \rangle = 1$, the collection $\{\mathbf{w}_1, \mathbf{v}_2\}$ is mutually orthogonal. Let $\mathbf{w}_2 = \frac{1}{|\mathbf{v}_2|}\mathbf{v}_2$, then $\{\mathbf{w}_1, \mathbf{w}_2\}$ is an orthonormal basis of a two-dimensional subspace contained in W. If W is two-dimensional, then it is the required basis for W. If not, we continue the process inductively, i.e., suppose that $\{\mathbf{w}_1, \ldots, \mathbf{w}_k\}$ is an orthonormal basis for a k-dimensional subspace of W. If $k < \dim W = r$, there exists a vector $\mathbf{v}$ not in the linear span of $\{\mathbf{w}_1, \ldots, \mathbf{w}_k\}$. Define

$$\mathbf{v}_{k+1} = \mathbf{v} - \sum_{j=1}^{k} \langle \mathbf{v}, \mathbf{w}_j \rangle \mathbf{w}_j.$$

Then $\mathbf{v}_{k+1}$ is orthogonal to the linear span of $\{\mathbf{w}_1, \ldots, \mathbf{w}_k\}$, linearly independent of $\{\mathbf{w}_1, \ldots, \mathbf{w}_k\}$ and contained in W. Define $\mathbf{w}_{k+1} = \frac{1}{|\mathbf{v}_{k+1}|}\mathbf{v}_{k+1}$. Then $\{\mathbf{w}_1, \ldots, \mathbf{w}_k, \mathbf{w}_{k+1}\}$ is an orthonormal basis of a $(k+1)$-dimensional subspace in W. Because the dimension of W is finite, this process will exhaust the dimension of W so that we obtain an orthonormal basis for W. ■

Given a vector subspace W in $\mathbb{R}^n$, we define the *orthogonal complement* of the subspace W as

$$W^{\perp} = \{\mathbf{v} \in \mathbb{R}^n : \langle \mathbf{v}, \mathbf{w} \rangle = 0, \quad \forall\, \mathbf{w} \in W\}. \tag{2.12}$$

Lemma 2.5.2 *If W is an r-dimensional vector subspace of $\mathbb{R}^n$, then $W^\perp$ is an $(n-r)$-dimensional vector subspace of $\mathbb{R}^n$ such that $W \cap W^\perp = \{\mathbf{0}\}$.*

Proof: We leave the proof that $W^\perp$ is a vector subspace as an exercise.

Note that if $\mathbf{w} \in W \cap W^\perp$, then it is in W and orthogonal to every vector in W. In particular, it is perpendicular to itself, i.e., $\langle \mathbf{w}, \mathbf{w} \rangle = 0$. This is possible only when $\mathbf{w} = \mathbf{0}$, which shows that $W \cap W^\perp = \{\mathbf{0}\}$.

According to the Gram-Schmidt process, there exists an orthonormal basis $\{\mathbf{w}_1, \dots, \mathbf{w}_r\}$ for the vector subspace W. We now continue the Gram-Schmidt process for $\mathbb{R}^n$ to get an orthonormal basis $\{\mathbf{w}_1, \dots, \mathbf{w}_r, \mathbf{w}_{r+1}, \dots, \mathbf{w}_n\}$ for $\mathbb{R}^n$. We denote $\mathbf{w}_{r+j}$ by $\mathbf{u}_j$ for $j = 1, \dots, n-r$.

By construction, the collection of vectors $\{\mathbf{u}_1, \dots, \mathbf{u}_{n-r}\}$ spans an $(n-r)$-dimensional vector subspace in $W^\perp$. If the dimension of $W^\perp$ is larger than $(n-r)$, there exists a vector $\mathbf{w}_\infty$ contained in $W^\perp$, such that $\{\mathbf{u}_1, \dots, \mathbf{u}_{n-r}, \mathbf{w}_\infty\}$ is a set of orthonormal vectors. Then $\mathbf{w}_\infty$ would not be in the linear span of $\{\mathbf{w}_1, \dots, \mathbf{w}_r, \mathbf{u}_1, \dots, \mathbf{u}_{n-r}\}$. This is a contradiction because $\mathbb{R}^n$ is n-dimensional. ■

Corollary 2.5.3 *Let W be a vector subspace of $\mathbb{R}^n$, and $W^\perp$ be its orthogonal complement. For each vector $\mathbf{v}$ in $\mathbb{R}^n$, there exists a unique vector $\mathbf{w}$ in W and $\mathbf{u}$ in $W^\perp$, such that $\mathbf{v} = \mathbf{w} + \mathbf{u}$.*

Proof: As noted in the proof of the last Lemma, there exists an orthonormal basis $\{\mathbf{w}_1, \dots, \mathbf{w}_r, \mathbf{u}_1, \dots, \mathbf{u}_{n-r}\}$ for $\mathbb{R}^n$, such that $\{\mathbf{w}_1, \dots, \mathbf{w}_r\}$ spans W and $\{\mathbf{u}_1, \dots, \mathbf{u}_{n-r}\}$ spans $W^\perp$. Every vector $\mathbf{v}$ is uniquely presented by a linear combination

$$\mathbf{v} = c_1\mathbf{w}_1 + \cdots + c_r\mathbf{w}_r + \lambda_1\mathbf{u}_1 + \cdots + \lambda_{n-r}\mathbf{u}_{n-r}.$$

Then

$$\mathbf{w} = c_1\mathbf{w}_1 + \cdots + c_r\mathbf{w}_r, \quad \mathbf{u} = \lambda_1\mathbf{u}_1 + \cdots + \lambda_{n-r}\mathbf{u}_{n-r}$$

are the two required vectors in the corollary.

Note that if there exists a different pair $\mathbf{w}' \in W$ and $\mathbf{u}' \in W^\perp$ such that $\mathbf{w}' + \mathbf{u}' = \mathbf{v} = \mathbf{w} + \mathbf{u}$, then

$$\mathbf{w}' - \mathbf{w} = \mathbf{u} - \mathbf{u}'.$$

It shows that the vectors $\mathbf{w}' - \mathbf{w}$ and $\mathbf{u} - \mathbf{u}'$ are both in $W \cap W^\perp$. Because this intersection contains only the zero vector $\mathbf{0}$, it follows that $\mathbf{w} = \mathbf{w}'$ and $\mathbf{u} = \mathbf{u}'$. ■

Exercise 2.5.4 Prove that $W^\perp$ is a vector subspace in $\mathbb{R}^n$.

Exercise 2.5.5 Show that if $\mathbf{A}$ is a conformal transformation and $\{\mathbf{v}_1, \dots, \mathbf{v}_n\}$ is an orthogonal basis of $\mathbb{R}^n$, then the same is true for $\{\mathbf{A}\mathbf{v}_1, \dots, \mathbf{A}\mathbf{v}_n\}$.

2.6 Orthogonal projections

Given a vector subspace W in $\mathbb{R}^n$, for any vector $\mathbf{v} \in \mathbb{R}^n$, let $\mathbf{w}$ and $\mathbf{u}$ be vectors in W and $W^\perp$, such that $\mathbf{v} = \mathbf{w} + \mathbf{u}$. The *orthogonal projection* of $\mathbf{v}$ onto the subspace W is the vector $\mathbf{w}$. Denote the map $\mathbf{v} \mapsto \mathbf{w}$ by $\mathbf{H}$ so that $\mathbf{Hv} = \mathbf{w}$.

By construction, the map $\mathbf{H}$ is a linear map from $\mathbb{R}^n$ to $\mathbb{R}^n$. The construction also shows that

$$\text{Image}\,\mathbf{H} = W, \quad \ker \mathbf{H} = W^\perp. \tag{2.13}$$

Therefore, the linear map $\mathbf{H}$ has two eigenvalues, namely "0" and "1". The subspace W is the linear span of all eigenvectors with the eigenvalue "1". The subspace $W^\perp$ is the linear span of all eigenvectors with the eigenvalue "0".

Next, we construct the matrix representation for $\mathbf{H}$ whenever a basis $\{\mathbf{w}_1, \ldots, \mathbf{w}_r\}$ for W is given. Each vector $\mathbf{w}_j$, $1 \leqslant j \leqslant r$ is a vector in $\mathbb{R}^n$. Consider the matrix whose j-th column is $\mathbf{w}_j$. Denote it by $\mathbf{W}$,

$$\mathbf{W} = (\mathbf{w}_1, \ldots, \mathbf{w}_r).$$

This is an $n \times r$-matrix. A vector $\mathbf{w}$ is in W if and only if there exist constants $c_1, \ldots, c_r$, such that $\mathbf{w} = c_1\mathbf{w}_1 + \cdots + c_r\mathbf{w}_r$. In terms of the matrix above, it is equivalent to

$$\mathbf{w} = (\mathbf{w}_1, \ldots, \mathbf{w}_r) \begin{pmatrix} c_1 \\ \vdots \\ c_r \end{pmatrix} = \mathbf{Wc},$$

where $\mathbf{c}$ is the vector $(c_1, \ldots, c_r)^T$ in $\mathbb{R}^r$.

Now, given any $\mathbf{v}$ in $\mathbb{R}^n$, $\mathbf{Hv}$ is in W. There must be a vector $\mathbf{c}$ in $\mathbb{R}^r$, such that $\mathbf{Hv} = \mathbf{Wc}$. We need to solve for $\mathbf{c}$ in terms of $\mathbf{v}$ and $\mathbf{W}$. According to Lemma 2.5.2, $\mathbf{v} - \mathbf{Wc}$ is orthogonal to every vector in W. Therefore,

$$\langle \mathbf{v} - \mathbf{Wc}, \mathbf{Wa} \rangle = 0$$

for any $\mathbf{a} = (a_1, \ldots, a_r)^T$. That is

$$0 = \langle \mathbf{v} - \mathbf{Wc}, \mathbf{Wa} \rangle = \mathbf{a}^T\mathbf{W}^T(\mathbf{v} - \mathbf{Wc}) = \langle \mathbf{a}, \mathbf{W}^T\mathbf{v} - \mathbf{W}^T\mathbf{Wc} \rangle,$$

where the inner product on the right-hand side is in the space $\mathbb{R}^r$. Because it holds for all $\mathbf{a} \in \mathbb{R}^r$, it is possible only when

$$\mathbf{W}^T\mathbf{Wc} = \mathbf{W}^T\mathbf{v}. \tag{2.14}$$

However, $\mathbf{W}^T\mathbf{W}$ is invertible because if there is $\mathbf{c}$ such that $\mathbf{W}^T\mathbf{Wc} = \mathbf{0}$, then for any $\mathbf{a}$ in $\mathbb{R}^r$,

$$0 = \langle \mathbf{a}, \mathbf{W}^T\mathbf{Wc} \rangle = \mathbf{a}^T\mathbf{W}^T\mathbf{Wc} = \langle \mathbf{Wa}, \mathbf{Wc} \rangle.$$

In particular, $\mathbf{Wc}$ is orthogonal to every column vector in the matrix $\mathbf{W}$. It follows that $\mathbf{Wc}$ is orthogonal to every vector in a basis. This is possible only when $\mathbf{c} = \mathbf{0}$. Therefore, $\mathbf{W}^T\mathbf{W}$ is invertible and the identity (2.14) above can be solved uniquely by

$$\mathbf{c} = (\mathbf{W}^T\mathbf{W})^{-1}\mathbf{W}^T\mathbf{v}.$$

Then, $\mathbf{Hv} = \mathbf{Wc} = \mathbf{W}(\mathbf{W}^T\mathbf{W})^{-1}\mathbf{W}^T\mathbf{v}$.

Theorem 2.6.1 *Let $\{\mathbf{w}_1, \ldots, \mathbf{w}_r\}$ be a basis for an r-dimensional vector subspace in $\mathbb{R}^n$. Let $\mathbf{W}$ be the $n \times r$-matrix whose columns are the given vectors. Then, the orthogonal projection on the linear spans of these column vectors is the matrix*

$$\mathbf{H} = \mathbf{W}(\mathbf{W}^T\mathbf{W})^{-1}\mathbf{W}^T. \tag{2.15}$$

It follows easily that

$$\mathbf{H}^2 = \mathbf{W}(\mathbf{W}^T\mathbf{W})^{-1}\mathbf{W}^T\mathbf{W}(\mathbf{W}^T\mathbf{W})^{-1}\mathbf{W}^T = \mathbf{W}(\mathbf{W}^T\mathbf{W})^{-1}\mathbf{W}^T = \mathbf{H}.$$

Note that if $\mathbf{A}$ is an invertible matrix, then $\mathbf{A}^T$ is also invertible and $(\mathbf{A}^T)^{-1} = (\mathbf{A}^{-1})^T$. Applying this identity to the matrix $\mathbf{H}^T$, we find that

$$\mathbf{H}^T = (\mathbf{W}(\mathbf{W}^T\mathbf{W})^{-1}\mathbf{W}^T)^T = \mathbf{W}((\mathbf{W}^T\mathbf{W})^{-1})^T\mathbf{W}^T = \mathbf{W}(\mathbf{W}^T\mathbf{W})^{-1}\mathbf{W}^T = \mathbf{H}.$$

Therefore, projection matrices are *idempotent*: $\mathbf{H}^2 = \mathbf{H}$ and *symmetric*: $\mathbf{H}^T = \mathbf{H}$. The following shows that the converse also holds.

Proposition 2.6.2 *If an $n \times n$-matrix $\mathbf{H}$ is idempotent and symmetric, then it is a projection matrix.*

Proof: Let $W =$ Image $\mathbf{H}$, the image space of the linear map $\mathbf{H}$. It is equal to the linear span of the column vectors of the matrix of $\mathbf{H}$. To prove the proposition, it suffices to prove that (a) for any $\mathbf{v} \in \mathbb{R}^n$, $\mathbf{v} - \mathbf{Hv}$ is orthogonal to W and that (b) $\ker \mathbf{H} = \{\mathbf{v} - \mathbf{Hv} : \mathbf{v} \in \mathbb{R}^n\}$ because combining these two statements combined implies that $W^\perp = \ker \mathbf{H}$.

To see (a), note that for all $\mathbf{x} \in \mathbb{R}^n$,

$$\langle \mathbf{Hx}, \mathbf{v} - \mathbf{Hv} \rangle = \mathbf{x}^T\mathbf{H}^T(\mathbf{v} - \mathbf{Hv}) = \mathbf{x}^T\mathbf{H}(\mathbf{v} - \mathbf{Hv}) = \mathbf{x}^T(\mathbf{Hv} - \mathbf{Hv}) = 0.$$

To see (b), note that

$$\mathbf{H}(\mathbf{v} - \mathbf{Hv}) = \mathbf{Hv} - \mathbf{H}^2\mathbf{v} = \mathbf{Hv} - \mathbf{Hv} = 0.$$

The proof is now completed. ■

Suppose that $\mathbf{H}$ is a projection matrix. Consider $\mathbf{I} - \mathbf{H}$. Because

$$(\mathbf{I} - \mathbf{H})(\mathbf{I} - \mathbf{H}) = \mathbf{I} - 2\mathbf{H} + \mathbf{H}^2 = \mathbf{I} - 2\mathbf{H} + \mathbf{H} = \mathbf{I} - \mathbf{H},$$

and

$$(\mathbf{I}-\mathbf{H})^T = \mathbf{I}^T - \mathbf{H}^T = \mathbf{I}-\mathbf{H},$$

According to the last proposition, $\mathbf{I}-\mathbf{H}$ is a projection matrix.

Corollary 2.6.3 *Suppose that* $\mathbf{H}$ *is the projection matrix whose image is the space* W*, then* $\mathbf{I}-\mathbf{H}$ *is the projection matrix of the orthogonal complement* $W^{\perp}$ *of* W.

Proof: If $\mathbf{w} \in W$, then $\mathbf{Hw} = \mathbf{w}$. Therefore, $(\mathbf{I}-\mathbf{H})\mathbf{w} = 0$ and hence $W \subseteq \ker(\mathbf{I}-\mathbf{H})$. On the other hand, if $\mathbf{v} \in \ker \mathbf{H}$, then $(\mathbf{I}-\mathbf{H})\mathbf{v} = \mathbf{v}$. Therefore, $W^{\perp}$ is the space of the eigenvectors of $\mathbf{I}-\mathbf{H}$ with eigenvalue 1. Because $W \cap W^{\perp} = \{\mathbf{0}\}$, it follows that $W = \ker(\mathbf{I}-\mathbf{H})$ and $W^{\perp} =$ Image $(\mathbf{I}-\mathbf{H})$. ■

Corollary 2.6.4 *Let* $\mathbf{H}$ *be an* $n \times n$ *projection matrix with rank* r. *Let* h_{ij} *be its* (i,j)*-th entry. Then,* $0 \leqslant h_{ii} \leqslant 1$ *for all* $1 \leqslant i \leqslant n$*, and* $\sum_{i=1}^{n} h_{ii} = r$.

Proof: Because $\mathbf{H}^2 = \mathbf{H}$, for each $1 \leqslant j, l \leqslant n$,

$$\sum_{j=1}^{n} h_{ij}h_{jk} = h_{ik}. \tag{2.16}$$

Because $h_{ij} = h_{ji}$ for all i, j, the above identity implies that

$$h_{ii} = \sum_{j=1}^{n} h_{ij}h_{ji} = \sum_{j=1}^{n} h_{ij}^2 = h_{ii}^2 + \sum_{j \neq i} h_{ij}^2 \geqslant 0. \tag{2.17}$$

Moreover, the same identity implies that

$$h_{ii}(1-h_{ii}) = \sum_{j \neq i} h_{ij}^2 \geqslant 0.$$

Because $h_{ii} \geqslant 0$, we also have $1 \geqslant h_{ii}$.

Since $\mathbf{H}$ is a projection matrix, its eigenvalues are either "1" or "0". Therefore the rank of $\mathbf{H}$ is equal to the trace of H. It follows that $\sum_{i=1}^{n} h_{ii} = r$.

■

Exercise 2.6.5 Considering vectors in a vector subspace W in $\mathbb{R}^n$ as vectors in $\mathbb{R}^n$, prove that the projection map $\mathbf{Hv} = \mathbf{w}$ as given in Section 2.6 is a linear map from $\mathbb{R}^n$ to $\mathbb{R}^n$.

2.7 Areas and volumes

Using elementary trigonometry and given a pair of vectors in $\mathbb{R}^2$:

$$\mathbf{v}_1 = \begin{pmatrix} v_{11} \\ v_{21} \end{pmatrix}, \quad \mathbf{v}_2 = \begin{pmatrix} v_{12} \\ v_{22} \end{pmatrix},$$

the area $\mathrm{Vol}(\mathbf{v}_1, \mathbf{v}_2)$ of the parallelogram spanned by them is equal to $|\mathbf{v}_1||\mathbf{v}_2|\sin\theta$, where θ is the angle between the two given vectors. Because $\cos\theta = \frac{\langle \mathbf{v}_1, \mathbf{v}_2\rangle}{|\mathbf{v}_1||\mathbf{v}_2|}$,

$$\begin{aligned}
\mathrm{Vol}(\mathbf{v}_1, \mathbf{v}_2)^2 &= |\mathbf{v}_1|^2|\mathbf{v}_2|^2\sin^2\theta = |\mathbf{v}_1|^2|\mathbf{v}_2|^2(1-\cos^2\theta) \\
&= |\mathbf{v}_1|^2|\mathbf{v}_2|^2 - \langle \mathbf{v}_1, \mathbf{v}_2\rangle^2 \qquad (2.18)\\
&= (v_{11}^2 + v_{21}^2)(v_{12}^2 + v_{22}^2) - (v_{11}v_{12} + v_{21}v_{22})^2 \\
&= (v_{11}v_{22} - v_{21}v_{12})^2. \qquad (2.19)
\end{aligned}$$

This observation demonstrates ways to compute the area without trigonometry. Consider the matrix $\mathbf{V}$ whose columns are the given pair of vectors in $\mathbb{R}^2$:

$$\mathbf{V} = (\mathbf{v}_1, \mathbf{v}_2) = \begin{pmatrix} v_{11} & v_{12} \\ v_{21} & v_{22} \end{pmatrix},$$

then

$$\mathbf{V}^T\mathbf{V} = \begin{pmatrix} \mathbf{v}_1^T \\ \mathbf{v}_2^T \end{pmatrix}(\mathbf{v}_1, \mathbf{v}_2) = \begin{pmatrix} \mathbf{v}_1^T\mathbf{v}_1 & \mathbf{v}_1^T\mathbf{v}_2 \\ \mathbf{v}_2^T\mathbf{v}_1 & \mathbf{v}_2^T\mathbf{v}_2 \end{pmatrix} = \begin{pmatrix} |\mathbf{v}_1|^2 & \langle \mathbf{v}_1, \mathbf{v}_2\rangle \\ \langle \mathbf{v}_2, \mathbf{v}_1\rangle & |\mathbf{v}_2|^2 \end{pmatrix}.$$

Therefore, there are two ways to compute Vol $(\mathbf{v}_1, \mathbf{v}_2)$, namely

$$\mathrm{Vol}(\mathbf{v}_1, \mathbf{v}_2) = |\det \mathbf{V}| \quad \text{and} \quad \mathrm{Vol}(\mathbf{v}_1, \mathbf{v}_2) = \sqrt{\det \mathbf{V}^T\mathbf{V}}. \qquad (2.20)$$

One of these expressions allows us to effectively generalize the definition of area to volume while the other provides us with an additional computation tool. In particular, given n-many linearly independent vectors $\mathbf{v}_1, \ldots, \mathbf{v}_n$ in $\mathbb{R}^n$, then the *volume of the parallelepiped* spanned by these vectors is the quantity

$$\mathrm{Vol}(\mathbf{v}_1, \ldots, \mathbf{v}_n) = |\det(\mathbf{v}_1, \ldots, \mathbf{v}_n)|. \qquad (2.21)$$

When we have n-many vectors in $\mathbb{R}^n$, the matrix $\mathbf{V} = (\mathbf{v}_1, \ldots, \mathbf{v}_n)$ is an $n \times n$-matrix. Therefore,

$$\det(\mathbf{V}^T\mathbf{V}) = \det \mathbf{V}^T \det \mathbf{V} = \det \mathbf{V} \det \mathbf{V} = \mathrm{Vol}(\mathbf{v}_1, \ldots, \mathbf{v}_n)^2. \qquad (2.22)$$

Thus, the identities in (2.20) remain valid.

Next, we consider a more general situation. Given two linearly independent vectors, $\mathbf{v}_1$ and $\mathbf{v}_2$, in $\mathbb{R}^3$, they span a parallelogram in $\mathbb{R}^3$. To compute its area, one could choose a unit-length vector $\mathbf{n}$ in the orthogonal complement of the linear span of the two given vectors. Then

$$\mathrm{Vol}(\mathbf{v}_1, \mathbf{v}_2) = \mathrm{Vol}(\mathbf{v}_1, \mathbf{v}_2, \mathbf{n}).$$

This situation can be generalized as having m-many linearly independent vectors $\mathbf{v}_1, \ldots, \mathbf{v}_m$ in $\mathbb{R}^{m+k}$, $(k > 0)$. Denote their linear span by V and the orthogonal complement by $V^{\perp}$. Let $\mathbf{n}_1, \ldots, \mathbf{n}_k$ be an orthonormal basis for $V^{\perp}$. We take this as a definition that

$$\mathrm{Vol}(\mathbf{v}_1, \ldots, \mathbf{v}_m) = \mathrm{Vol}(\mathbf{v}_1, \ldots, \mathbf{v}_m, \mathbf{n}_1, \ldots, \mathbf{n}_k). \tag{2.23}$$

The next observation allows us to compute the volume of the parallelepiped effectively, without explicitly involving of its orthogonal complement.

Proposition 2.7.1 *Let $\{\mathbf{v}_1, \ldots, \mathbf{v}_m\}$ be a collection of linearly independent vectors in $\mathbb{R}^{m+k}$. Then*

$$\mathrm{Vol}(\mathbf{v}_1, \ldots, \mathbf{v}_m) = \sqrt{\det \mathbf{V}^T\mathbf{V}}$$

where $\mathbf{V} = (\mathbf{v}_1, \ldots, \mathbf{v}_m)$.

Proof: Let $\mathbf{N} = (\mathbf{n}_1, \ldots, \mathbf{n}_k)$ be the matrix of the vectors in the orthogonal complement $V^{\perp}$. Then, $(\mathbf{V}, \mathbf{N})$ is an $(m+k) \times (m+k)$-matrix. Apply formula (2.22),

$$\begin{aligned}\mathrm{Vol}(\mathbf{v}_1, \ldots, \mathbf{v}_m, \mathbf{n}_1, \ldots, \mathbf{n}_k)^2 &= \det\left((\mathbf{V}, \mathbf{N})^T(\mathbf{V}, \mathbf{N})\right) \\ &= \det\begin{pmatrix} \mathbf{V}^T\mathbf{V} & \mathbf{V}^T\mathbf{N} \\ \mathbf{N}^T\mathbf{V} & \mathbf{N}^T\mathbf{N} \end{pmatrix}.\end{aligned}$$

Because $\{\mathbf{n}_1, \ldots, \mathbf{n}_k\}$ is an orthonormal basis for $V^{\perp}$, $\mathbf{N}^T\mathbf{N} = \mathbf{I}$, the $k \times k$-identity matrix. Because these vectors are all orthogonal to those in $\{\mathbf{v}_1, \ldots, \mathbf{v}_m\}$, $\mathbf{N}^T\mathbf{V}$ and $\mathbf{V}^T\mathbf{N}$ are both zero matrices. Therefore,

$$\det\begin{pmatrix} \mathbf{V}^T\mathbf{V} & \mathbf{V}^T\mathbf{N} \\ \mathbf{N}^T\mathbf{V} & \mathbf{N}^T\mathbf{N} \end{pmatrix} = \det\begin{pmatrix} \mathbf{V}^T\mathbf{V} & \mathbf{0} \\ \mathbf{0} & \mathbf{I} \end{pmatrix} = \det \mathbf{V}^T\mathbf{V}.$$

Then, the proposition is a consequence of (2.23). ■

Exercise 2.7.2 Given two vectors $\mathbf{v}_1$, $\mathbf{v}_2$ in $\mathbb{R}^n$ for $n \geqslant 2$, show that the area of the parallelogram spanned by them is equal to $\sqrt{|\mathbf{v}_1|^2|\mathbf{v}_2|^2 - \langle \mathbf{v}_1, \mathbf{v}_2\rangle}$.

Exercise 2.7.3 Given two pairs of vectors $\{\mathbf{v}_1, \mathbf{v}_2\}$, $\{\mathbf{w}_1, \mathbf{w}_2\}$ in $\mathbb{R}^n$ for $n \geqslant 2$, suppose that they are linearly dependent, i.e., there are numbers a, b, c, d, such that $\mathbf{v}_1 = a\mathbf{w}_1 + b\mathbf{w}_2$ and $\mathbf{v}_2 = c\mathbf{w}_1 + d\mathbf{w}_2$. Show that

$$\mathrm{Vol}(\mathbf{v}_1, \mathbf{v}_2) = \left|\det\begin{pmatrix} a & b \\ c & d \end{pmatrix}\right| \mathrm{Vol}(\mathbf{w}_1, \mathbf{w}_2).$$

Exercise 2.7.4 Let $\{\mathbf{v}_1, \ldots, \mathbf{v}_m\}$ be a collection of linearly independent vectors in $\mathbb{R}^{m+k}$ and $\mathbf{A} \in O(m+k)$. Show that

$$\mathrm{Vol}(\mathbf{v}_1, \ldots, \mathbf{v}_m) = \mathrm{Vol}(\mathbf{A}\mathbf{v}_1, \ldots, \mathbf{A}\mathbf{v}_m).$$

Exercise 2.7.5 Let $\{\mathbf{v}_1, \ldots, \mathbf{v}_m\}$ be a collection of linearly independent vectors in $\mathbb{R}^{m+k}$ and $\mathbf{A} \in CO(m+k)$. Derive a relation between $\mathrm{Vol}(\mathbf{v}_1, \ldots, \mathbf{v}_m)$ and $\mathrm{Vol}(\mathbf{A}\mathbf{v}_1, \ldots, \mathbf{A}\mathbf{v}_m)$.

Chapter 3

Geometry of Graphs

3.1 Graphs in Euclidean spaces

Suppose that $f(\mathbf{x})$ is a real-valued function defined in an open subset U of $\mathbb{R}^n$, then the graph G:

$$\{\mathbf{y} \in \mathbb{R}^{n+1} : y_j = x_j \text{ for } 1 \leqslant j \leqslant n, y_{n+1} = f(\mathbf{x}), \mathbf{x} \in U\}, \tag{3.1}$$

is a set of points in $\mathbb{R}^{n+1}$.

Analytically, we describe this set as the image of a function $\phi : U \to \mathbb{R}^{n+1}$. Explicitly, it is given by the function

$$\phi(\mathbf{x}) = \begin{pmatrix} \mathbf{x} \\ f(\mathbf{x}) \end{pmatrix} = \begin{pmatrix} x_1 \\ \vdots \\ x_n \\ f(\mathbf{x}) \end{pmatrix}. \tag{3.2}$$

The function ϕ represents the graph analytically, while the set $G = \phi(U)$ in $\mathbb{R}^{n+1}$ represents the graph geometrically. We will address both ϕ and G as the graph of f.

The differential of the map ϕ is the following matrix,

$$d\phi = \begin{pmatrix} 1 & 0 & \cdots & 0 \\ 0 & 1 & \cdots & 0 \\ \vdots & \vdots & & \vdots \\ 0 & 0 & \cdots & 1 \\ \dfrac{\partial f}{\partial x_1} & \dfrac{\partial f}{\partial x_2} & \cdots & \dfrac{\partial f}{\partial x_n} \end{pmatrix} = \begin{pmatrix} \mathbf{I} \\ df \end{pmatrix}, \tag{3.3}$$

where df is the differential of the function f. The image of a tangent vector $\mathbf{v}$ via the differential of the map ϕ is

$$d\phi(\mathbf{v}) = \begin{pmatrix} \mathbf{I} \\ df \end{pmatrix} \begin{pmatrix} v_1 \\ \vdots \\ v_n \end{pmatrix} = \begin{pmatrix} \mathbf{v} \\ df(\mathbf{v}) \end{pmatrix} = \begin{pmatrix} \mathbf{v} \\ \sum_{k=1}^{n} \frac{\partial f}{\partial x_k} v_k \end{pmatrix}, \tag{3.4}$$

which is also equal to

$$\begin{pmatrix} \mathbf{v} \\ \langle \nabla_f, \mathbf{v} \rangle \end{pmatrix}. \tag{3.5}$$

For instance, the image of a basic vector $\mathbf{e}_j$ in $\mathbb{R}^n$ is equal to

$$d\phi(\mathbf{e}_j) = \begin{pmatrix} \mathbf{e}_j \\ \frac{\partial f}{\partial x_j} \end{pmatrix}. \tag{3.6}$$

Given a point $\phi(\mathbf{x})$ on the graph G of f, the *tangent space of the graph* G at $\phi(\mathbf{x})$ is the set

$$T_{\phi(\mathbf{x})}G = \{\mathbf{w} = d\phi(\mathbf{v}) : \mathbf{v} \in \mathbb{R}^n\}. \tag{3.7}$$

Normal vectors of the graph G at $\phi(\mathbf{x})$ are vectors $\mathbf{n}$ such that $\langle \mathbf{w}, \mathbf{n} \rangle = 0$ for all $\mathbf{w} \in T_{\phi(\mathbf{x})}G$.

Consider the vector

$$\mathbf{n} = \begin{pmatrix} -\nabla_f \\ 1 \end{pmatrix}, \tag{3.8}$$

where the gradient is evaluated at the point $\mathbf{x}_0$. Because

$$\langle \mathbf{n}, d\phi(\mathbf{e}_j) \rangle = -\langle \nabla_f, \mathbf{e}_j \rangle + \frac{\partial f}{\partial x_j} = 0,$$

and the collection of vectors $\{d\phi(\mathbf{e}_1), \ldots, d\phi(\mathbf{e}_n)\}$ spans the tangent space of the graph of the map ϕ at $\mathbf{x}_0$, the vector $\mathbf{n}$ is a normal vector. In the future, we address the vectors $\mathbf{n}$ and $\frac{\mathbf{n}}{|\mathbf{n}|}$ as upward normal vectors. They are conceived as pointing upward because their last coordinate is positive.

Proposition 3.1.1 *Suppose that $f(\mathbf{x})$ is a differentiable function defined on $U \subseteq \mathbb{R}^n$. If $\mathbf{x}_0$ in U is a critical point of f, then the differential of the graph of the function f identifies the tangent space of $\mathbb{R}^n$ at the point $\mathbf{x}_0$ with the vector subspace*

$$\{(w_1, \ldots, w_n, w_{n+1})^T \in \mathbb{R}^{n+1} : w_{n+1} = 0\}.$$

In other words, it is the standard embedding of $\mathbb{R}^n$ in $\mathbb{R}^{n+1}$.

Proof: If $\mathbf{x}_0$ is a critical point of the function $f(\mathbf{x})$, the gradient vector of f at $\mathbf{x}_0$ is equal to zero. According to (3.3), the differential of ϕ at $\mathbf{x}_0$ is simply the matrix (1.10). In such a situation, based on (3.8), the upward normal vector is the basic vector $\mathbf{e}_{n+1}$. ■

According to the observation in the last proposition, we often simply denote $d\phi(\mathbf{e}_j)$ by $\mathbf{e}_j$ for $j = 1, \ldots, n$ when $\mathbf{x}_0$ is a critical point of f.

We may even assume that $f(\mathbf{x}_0) = \mathbf{0}$. If $f(\mathbf{x}_0) \neq \mathbf{0}$, a translation in $\mathbb{R}^{n+1}$ given by the vector $-\phi(\mathbf{x}_0)$ in $\mathbb{R}^{n+1}$ moves the graph from $\phi(\mathbf{x}_0)$ to the origin $\mathbf{0}$ of $\mathbb{R}^{n+1}$. Strictly speaking, it is the graph of the function $f(\mathbf{x}) - f(\mathbf{x}_0)$ with respect to the variable $\mathbf{x}$.

One more change of variable $\mathbf{x} = \mathbf{x}_0 + \mathbf{t}$, and the function $f(\mathbf{x}) - f(\mathbf{x}_0) = f(\mathbf{x}_0+\mathbf{t}) - f(\mathbf{x}_0)$ becomes a function of the variable $\mathbf{t}$, such that when $\mathbf{t} = \mathbf{0}$, the value of the given function is equal to zero. Therefore, we assume that $f(\mathbf{0}) = \mathbf{0}$, and the graph of f goes through the origin of $\mathbb{R}^{n+1}$. So does the tangent space of the graph of the function f, $T_{\phi(\mathbf{0})}G$. According to the Gram-Schmidt Process in Lemma 2.5.1, there exists an orthonormal basis $\{\mathbf{v}_1, \ldots, \mathbf{v}_n\}$ in $T_{\phi(\mathbf{0})}G$. Let $\mathbf{n}$ be the normal vector of the graph at $\phi(\mathbf{0})$. Let $\mathbf{A}$ be the matrix whose columns are

$$\left\{\mathbf{v}_1, \ldots, \mathbf{v}_n, \frac{1}{|\mathbf{n}|}\mathbf{n}\right\}.$$

Equivalently, $\mathbf{A}\mathbf{e}_j = \mathbf{v}_j$ for $1 \leqslant j \leqslant n$ and $\mathbf{A}\mathbf{e}_{n+1} = \frac{1}{|\mathbf{n}|}\mathbf{n}$. Therefore, $\mathbf{A}^{-1}$ sends the tangent space $T_{\phi(\mathbf{x}_0)}G$ to the standard embedding of $\mathbb{R}^n$ into $\mathbb{R}^{n+1}$, such that the upward normal vector of the tangent space is sent to the upward normal of the standard embedding.

Therefore, whenever we want to study the property of the graph of a function at one point, we can assume that the point in question takes value at $\mathbf{0}$ and it is a critical point. Because translations and orthogonal transformations are rigid motions, the geometry on the graph does not change. Therefore, we will often simplify our analysis by moving our graphs through rigid motion, so that the point in question is a critical point.

3.2 Normal sections

To describe the geometry of a graph, we can literally look at it from multiple perspectives. We first analyze the property of the function $f(\mathbf{x})$ with respect to the variable $\mathbf{x} = (x_1, \ldots, x_n)^T$ by examining the property of f one coordinate at a time. For example, if one is interested in the behavior of the function f near the point $\mathbf{x}$ with respect to x_j, then one considers the function with a single variable t:

$$f_j(t) = f(x_1, \ldots, x_{j-1}, t + x_j, x_{j+1}, \ldots, x_n), \tag{3.9}$$

and t varies in a small neighborhood of 0, i.e., $t \in (-\epsilon, \epsilon)$ for some $\epsilon > 0$. The graph of the function f_j is now in $\mathbb{R}^2$ and could be considered a path as defined by the function

$$\phi_j(t) = \begin{pmatrix} t \\ f_j(t) \end{pmatrix}. \tag{3.10}$$

More generally, if $\mathbf{v} = (v_1, \ldots, v_n)^T$ is a vector in $\mathbb{R}^n$, then we can consider the function of a single variable, $f(\mathbf{x} + t\mathbf{v})$. The graph of this function is contained in a two-dimensional space.

Alternatively, we can consider the linear span of $d\phi(\mathbf{v})$ and the normal vector $\mathbf{n}$. Their linear span is a two-dimensional space that, ideally, cuts across the graph of the function f along a smooth curve. To be more precise, let $\mathbf{w}$ be a vector in $\mathbb{R}^{n+1}$, tangent to the graph at $\phi(\mathbf{x})$. Let

$$N(\mathbf{w}) = \{\mathbf{y} \in \mathbb{R}^{n+1} : \mathbf{y} = \phi(\mathbf{x}) + \mu\mathbf{w} + \nu\mathbf{n} \text{ for some } \mu, \nu \in \mathbb{R}\}.$$

Then the *normal section* in the direction of $\mathbf{w}$ at the point $\phi(\mathbf{x})$ is the set $G(\mathbf{w}) = \phi(U) \cap N(\mathbf{w})$.

Note that $N(\mathbf{w})$ is a plane in $\mathbb{R}^{n+1}$ through the point $\phi(x)$ and spanned by $\mathbf{w}$ and $\mathbf{n}$. The set $G(\mathbf{w})$ is a curve in the plane of $N(\mathbf{w})$ and passes through the point $\phi(\mathbf{x})$. Because $\mathbf{n}$ is perpendicular to every vector tangent to the graph at the point $\phi(\mathbf{x})$, the tangent vector to the curve $G(\mathbf{w})$ is in the linear span of $\mathbf{w}$. To further illustrate this construction, we focus on a situation vital to our future analysis.

Suppose that $\mathbf{x}_0$ is a critical point of the function f. Then, the normal vector $\mathbf{n}$ is the basic vector $\mathbf{e}_{n+1}$ and the tangent space of the graph at $\phi(\mathbf{x}_0)$ is the standard embedding of $\mathbb{R}^n$ into $\mathbb{R}^{n+1}$.

Let $\mathbf{x} = \mathbf{x}_0 + t\mathbf{e}_j$ be a line in the domain U of the function f. If we evaluate the function f along this line, then $f(\mathbf{x}) = f_j(t)$ as given in (3.9). Let $\nu(t) = f_j(t) - f(\mathbf{x}_0)$. By assumption on $\mathbf{x}_0$, $t = 0$ is a critical point of $f_j(t)$. Therefore, $\nu(0) = 0$ and $\nu'(0) = 0$. Now the graph of the function f restricted over the line $\mathbf{x} = \mathbf{x}_0 + t\mathbf{e}_j$ is equal to

$$\begin{aligned} \phi(\mathbf{x}_0 + t\mathbf{e}_j) = \begin{pmatrix} \mathbf{x}_0 + t\mathbf{e}_j \\ f_j(t) \end{pmatrix} &= \begin{pmatrix} \mathbf{x}_0 \\ f(\mathbf{x}_0) \end{pmatrix} + t\begin{pmatrix} \mathbf{e}_j \\ 0 \end{pmatrix} + \nu\begin{pmatrix} 0 \\ 1 \end{pmatrix} \\ &= \phi(\mathbf{x}_0) + td\phi(\mathbf{e}_j) + \nu\mathbf{n}(\mathbf{x}_0). \end{aligned}$$

Therefore, we recover the normal section of the graph of f at the critical point $\phi(\mathbf{x}_0)$ in the direction of $d\phi(\mathbf{e}_j)$, which is also the basic vector $\mathbf{e}_j$ of $\mathbb{R}^{n+1}$. It is now apparent that the tangent vector of $\phi(\mathbf{x}_0 + t\mathbf{e}_j)$ at $t = 0$ is exactly $\mathbf{e}_j$.

If $\mathbf{v}$ is a unit vector in $\mathbb{R}^n$, one can consider the line $\mathbf{x} = \mathbf{x}_0 + t\mathbf{v}$ in U. Evaluate the function f along this line, one gets a function of t. Let $\nu(t) = f(\mathbf{x}_0 + t\mathbf{v}) - f(\mathbf{x}_0)$, then $\nu(0) = 0$ and $\nu'(0) = 0$. This yields the normal section of the graph of f at the critical point $\mathbf{x}_0$ in the direction of $d\phi(\mathbf{v})$:

$$\phi(\mathbf{x}_0 + t\mathbf{v}) = \phi(\mathbf{x}_0) + td\phi(\mathbf{v}) + \nu\mathbf{n}. \tag{3.11}$$

Because $\nu'(0) = 0$, the tangent vector of this curve at $\phi(\mathbf{x}_0)$ is precisely $d\phi(\mathbf{v})$.

In the future, we will denote this normal section by $G(d\phi(\mathbf{v}))$ and $G(\mathbf{v})$. At a critical point, with the standard embedding of $\mathbb{R}^n$ into $\mathbb{R}^{n+1}$, there is little ambiguity in our notations.

3.3 Cross sections in high dimension

The concept of normal sections can be extended to a higher dimensions. Let $\mathbf{w}_1, \ldots, \mathbf{w}_k$ be vectors linearly independent in $\mathbb{R}^{n+1}$ and tangent to the graph at $\phi(\mathbf{x})$. Let $N(\mathbf{w}_1, \ldots, \mathbf{w}_k)$ be the set

$$\{\mathbf{y} \in \mathbb{R}^{n+1} : \mathbf{y} = \phi(\mathbf{x}) + \sum_{j=1}^{k} \mu_j \mathbf{w}_j + \nu \mathbf{n} \text{ for some } \mu_1, \ldots, \mu_k, \nu \in \mathbb{R}\}.$$

The set $G(\mathbf{w}_1, \ldots, \mathbf{w}_k) = \phi(U) \cap N(\mathbf{w}_1, \ldots, \mathbf{w}_k)$ is called a **k-dimensional cross section** at the point $\phi(\mathbf{x})$, or a k-dimensional cross section in the span of $\mathbf{w}_1, \ldots, \mathbf{w}_k$.

Whenever $\mathbf{w}$ is a linear combination of $\mathbf{w}_1$ and $\mathbf{w}_2$, then $G(\mathbf{w}) \subset G(\mathbf{w}_1, \mathbf{w}_2)$. When $\{\mathbf{w}_1, \ldots, \mathbf{w}_n\}$ forms a basis for the tangent space $T_{\phi(\mathbf{x})}G$, then $G = G(\mathbf{w}_1, \ldots, \mathbf{w}_n)$. More generally, we have a series of inclusions:

$$G(\mathbf{w}_i) \subset G(\mathbf{w}_i, \mathbf{w}_j) \subset G(\mathbf{w}_i, \mathbf{w}_j, \mathbf{w}_k) \subset \cdots \subset G(\mathbf{w}_1, \ldots, \mathbf{w}_n). \tag{3.12}$$

3.4 First fundamental forms

As seen in Chapter 2, the notion of inner product leads to the concepts of length, distance, angles, areas and volumes. However, when conceiving the graph of a function f on $\mathbb{R}^n$ as a map from $\mathbb{R}^n$ to $\mathbb{R}^{n+1}$, the length of a tangent vector $\mathbf{v}$ in $T_{\mathbf{x}}\mathbb{R}^n$ is not necessarily equal to the length of the vector $d\phi(\mathbf{v})$ in $T_{\phi(\mathbf{x})}\mathbb{R}^{n+1}$. Therefore, a measurement of the relation between $|\mathbf{v}|^2$ and $|d\phi(\mathbf{v})|^2$ is a reflection of the geometry of the graph G, which in turn reflects the property of the function f.

For example, let $\mathbf{v}$ be a tangent vector to $\mathbb{R}^n$ at a point $\mathbf{x}$. Its length in $\mathbb{R}^n$ with respect to the standard inner product is $|\mathbf{v}| = \sqrt{\langle \mathbf{v}, \mathbf{v} \rangle}$. If f is a function in an open set containing $\mathbf{x}$, then the image of $\mathbf{v}$ via the graph ϕ of f is

$$d\phi(\mathbf{v}) = \begin{pmatrix} \mathbf{v} \\ \langle \nabla_f, \mathbf{v} \rangle \end{pmatrix}.$$

Therefore, the length of the image vector is

$$|d\phi(\mathbf{v})| = \sqrt{\langle d\phi(\mathbf{v}), d\phi(\mathbf{v})\rangle} = \sqrt{|\mathbf{v}|^2 + \langle \nabla_f, \mathbf{v}\rangle^2}. \tag{3.13}$$

In general, if $\mathbf{v}$ and $\mathbf{w}$ are vectors in $\mathbb{R}^n$, then

$$\langle d\phi(\mathbf{v}), d\phi(\mathbf{w})\rangle = \mathbf{v}^T\mathbf{w} + \langle \nabla_f, \mathbf{v}\rangle\langle \nabla_f, \mathbf{w}\rangle = \langle \mathbf{v}, \mathbf{w}\rangle + \langle \nabla_f, \mathbf{v}\rangle\langle \nabla_f, \mathbf{w}\rangle. \tag{3.14}$$

Using the convention in (1.14), we find the above to be equal to

$$\mathbf{v}^T(\mathbf{I} + {\nabla_f}^T\nabla_f)\mathbf{w}. \tag{3.15}$$

Apparently, $\mathbf{I} + {\nabla_f}^T\nabla_f$ is a symmetric $n \times n$-matrix. Therefore, this matrix encodes all information for the inner products among the tangent vectors on the graph of f.

As a matter of definition, the *first fundamental form* $\mathbb{I}$ is the symmetric bilinear symmetric form $\mathbf{I} + {\nabla_f}^T\nabla_f$. In other words,

$$\mathbb{I}(\mathbf{v}, \mathbf{w}) = \mathbf{v}^T\mathbf{w} + \langle \nabla_f, \mathbf{v}\rangle\langle \nabla_f, \mathbf{w}\rangle. \tag{3.16}$$

In modern geometry, the first fundamental form is also known as the "induced" Riemannian metric of the graph of f. It is an "induced" object because the sense of distance between two vectors in $T_{\phi(\mathbf{x})}\mathbb{R}^{n+1}$ is due to the standard inner product in $\mathbb{R}^{n+1}$.

If $\mathbf{x}$ is a critical point of the function f, then $\nabla_f = \mathbf{0}$ at $\mathbf{x}$, and $\mathbb{I}(\mathbf{v}, \mathbf{w}) = \langle \mathbf{v}, \mathbf{w}\rangle$ for any $\mathbf{v}, \mathbf{w}$ in $\mathbb{R}^n$. Meanwhile, the tangent space of the graph at $\phi(\mathbf{x})$ is the standard embedding of $\mathbb{R}^n$ in $\mathbb{R}^{n+1}$. Therefore, the evaluation of the first fundamental form on a pair of tangent vectors at a critical point of the function f is simply the usual inner product in $\mathbb{R}^{n+1}$.

Chapter 4

Curvatures

If f is a function of a single variable x, then the graph of the function $f(x)$ is in the xy-plane such that a point in $\mathbb{R}^2$ is on the graph of f if and only if $y = f(x)$.

Suppose $f(0) = 0$ and it is a minimum, then $f'(0) = 0$ and $f''(0) > 0$. Geometrically, the graph of f passes through the origin of the xy-plane, it is tangent to the x-axis and is concave upward. The larger $f''(0)$, the more acute the concavity. Therefore, the quantity $f''(0)$ at a critical point of f is a useful numerical measure on the geometry of the curve at the critical point.

For example, the function $f(x) = 1 - \cos(kx)$ attains a minimum at $x = 0$. The graph of $f(x)$ near $x = 0$ is concave upward with the concavity increasing as k increases. In addition, $f''(0) = k^2$. Therefore, $f''(0)$ can be used to detect the magnitude of concavity. In the rest of this chapter, we use various generalizations and combinations of calculus methods to describe such concavity, also known as curvature.

4.1 Normal curvatures

As noted in Section 3.3, when a function f has $\mathbf{x}$ as a critical point, the tangent space of the graph of f is the standard embedding of $\mathbb{R}^n$ in $\mathbb{R}^{n+1}$. Furthermore, the concavity of the graph of f at the point $\phi(\mathbf{x})$ can be detected by studying the change in the unit upward normal vector $\mathbf{u} = \frac{1}{|\mathbf{n}|}\mathbf{n}$ with respect to various unit tangent vectors in $\mathbb{R}^n$.

4.1.1 Definition

A function $\mathbf{w} : \mathbb{R}^{n+1} \to \mathbb{R}^{n+1}$ is conceived as a vector field because for each point in $\mathbb{R}^{n+1}$, it assigns a vector. If $\mathbf{v}$ is a tangent vector of $\mathbb{R}^{n+1}$ and $\mathbf{w} = (w_1, \ldots, w_{n+1})^T$, the *directional derivative* of $\mathbf{w}$ with respect to $\mathbf{v}$, $\nabla_{\mathbf{v}}\mathbf{w}$ is defined as

$$\nabla_{\mathbf{v}}\mathbf{w} = (\langle \nabla w_1, \mathbf{v}\rangle, \ldots, \langle \nabla w_{n+1}, \mathbf{v}\rangle)^T. \tag{4.1}$$

This is a new vector field. Given this definition, it is easy to verify that if $\mathbf{w}$ and $\mathbf{y}$ are two vector fields, then the directional derivative satisfies a product rule:

$$\nabla_{\mathbf{v}}\langle \mathbf{w}, \mathbf{y}\rangle = \langle \nabla_{\mathbf{v}}\mathbf{w}, \mathbf{y}\rangle + \langle \mathbf{w}, \nabla_{\mathbf{v}}\mathbf{y}\rangle.$$

Now calculate $\nabla_{\mathbf{v}}\mathbf{u}$, the directional derivative of the upward pointing unit normal vector $\mathbf{u}$ with respect to a tangent vector $\mathbf{v}$ of the graph G. Because $\mathbf{u}$ has unit length,

$$\nabla_{\mathbf{v}}\langle \mathbf{u}, \mathbf{u}\rangle = \nabla_{\mathbf{v}}1 = 0.$$

According to product rule,

$$\langle \nabla_{\mathbf{v}}\mathbf{u}, \mathbf{u}\rangle + \langle \mathbf{u}, \nabla_{\mathbf{v}}\mathbf{u}\rangle = \nabla_{\mathbf{v}}\langle \mathbf{u}, \mathbf{u}\rangle = 0.$$

Because the inner product is symmetric, $2\langle \nabla_{\mathbf{v}}\mathbf{u}, \mathbf{u}\rangle = 0$. Therefore, the vector field $\nabla_{\mathbf{v}}\mathbf{u}$ is tangent to the graph G of the function f. Its component in the direction of $\mathbf{v}$ is captured by $\langle \nabla_{\mathbf{v}}\mathbf{u}, \mathbf{v}\rangle$.

To calculate this quantity effectively, we consider a more general concept. When $\mathbf{v}$ and $\mathbf{w}$ are tangent to the graph G at the same point,

$$\langle \nabla_{\mathbf{v}}\mathbf{u}, \mathbf{w}\rangle = \langle \nabla_{\mathbf{v}}\Big(\frac{1}{|\mathbf{n}|}\mathbf{n}\Big), \mathbf{w}\rangle = \langle \Big(\nabla_{\mathbf{v}}\frac{1}{|\mathbf{n}|}\Big)\mathbf{n} + \frac{1}{|\mathbf{n}|}\nabla_{\mathbf{v}}\mathbf{n}, \mathbf{w}\rangle.$$

Because $\langle \mathbf{n}, \mathbf{w}\rangle = 0$, the above is simplified to

$$\frac{1}{|\mathbf{n}|}\langle \nabla_{\mathbf{v}}\mathbf{n}, \mathbf{w}\rangle.$$

This quantity depends on both $\mathbf{v}$ and $\mathbf{w}$. The following

$$\Pi(\mathbf{v}, \mathbf{w}) = -\frac{1}{|\mathbf{n}|}\langle \nabla_{\mathbf{v}}\mathbf{n}, \mathbf{w}\rangle \tag{4.2}$$

is known as the *second fundamental form* of the graph G evaluated at the pair of vectors $\mathbf{v}$ and $\mathbf{w}$.

If $\mathbf{v} = d\phi(\mathbf{e}_j)$ and $\mathbf{w} = d\phi(\mathbf{e}_k)$, then according to (3.6)

$$d\phi(\mathbf{e}_j) = \begin{pmatrix} \mathbf{e}_j \\ \dfrac{\partial f}{\partial x_j} \end{pmatrix}, \quad d\phi(\mathbf{e}_k) = \begin{pmatrix} \mathbf{e}_k \\ \dfrac{\partial f}{\partial x_k} \end{pmatrix}. \tag{4.3}$$

Because $\mathbf{n} = (-\frac{\partial f}{\partial x_1}, \ldots, -\frac{\partial f}{\partial x_n}, 1)^T$, by the definition of the directional derivative

$$\nabla_{d\phi(\mathbf{e}_j)}\mathbf{n} = \begin{pmatrix} -\langle \nabla \frac{\partial f}{\partial x_1}, d\phi(\mathbf{e}_j)\rangle \\ \vdots \\ -\langle \nabla \frac{\partial f}{\partial x_n}, d\phi(\mathbf{e}_j)\rangle \\ 0 \end{pmatrix} = \begin{pmatrix} -\frac{\partial^2 f}{\partial x_j \partial x_1} \\ \vdots \\ -\frac{\partial^2 f}{\partial x_j \partial x_n} \\ 0 \end{pmatrix}. \tag{4.4}$$

Therefore,

$$\langle \nabla_{d\phi(\mathbf{e}_j)}\mathbf{n}, d\phi(\mathbf{e}_k)\rangle = -\frac{\partial^2 f}{\partial x_j \partial x_k}. \tag{4.5}$$

Considering that any pair of tangent vectors $\mathbf{v}$ and $\mathbf{w}$ of the graph G are linear combinations of $d\phi(\mathbf{e}_j)$s, and $\langle \nabla_{\mathbf{v}}\mathbf{u}, \mathbf{w}\rangle$ is linear in the variables $\mathbf{v}$ and $\mathbf{w}$, respectively, the above computation yields the identity

$$\mathrm{II}(\mathbf{v}, \mathbf{w}) = -\langle \nabla_{\mathbf{v}}\mathbf{u}, \mathbf{w}\rangle = -\frac{1}{|\mathbf{n}|}\langle \nabla_{\mathbf{v}}\mathbf{n}, \mathbf{w}\rangle = \frac{\mathbf{v}^T \ddot{\mathbf{F}} \mathbf{w}}{\sqrt{1 + |\nabla_f|^2}}. \tag{4.6}$$

The *normal curvature* of a graph at the point $\phi(\mathbf{x})$ in the direction of $\mathbf{v}$ is defined as the ratio

$$C_{\mathbf{v}} = \frac{\mathrm{II}(\mathbf{v}, \mathbf{v})}{\mathrm{I}(\mathbf{v}, \mathbf{v})}. \tag{4.7}$$

The above computation provides a computational formal, which we capture in the next observation.

Proposition 4.1.1 *The normal curvature $C_{\mathbf{v}}$ at a point $\phi(\mathbf{x})$ of the graph in the direction of $\mathbf{v}$ is given by*

$$C_{\mathbf{v}} = \frac{\mathrm{II}(\mathbf{v}, \mathbf{v})}{\mathrm{I}(\mathbf{v}, \mathbf{v})} = \frac{\mathbf{v}^T \ddot{\mathbf{F}} \mathbf{v}}{(|\mathbf{v}|^2 + |\nabla_{\mathbf{v}} f|^2)\sqrt{1 + |\nabla_f|^2}}. \tag{4.8}$$

Allowing $\mathbf{v}$ in the above definition to vary in $T_{\phi(\mathbf{x})}G$, we obtain a function

$$C : T_{\phi(\mathbf{x})}G \to \mathbb{R}. \tag{4.9}$$

This function is the *normal curvature function* of the graph of the function f at the point $\mathbf{x}$.

4.1.2 Principal curvatures and principal directions

Note that if a is a non-zero constant, based on the definition given in (4.7),

$$C_{a\mathbf{v}} = C_{\mathbf{v}}. \tag{4.10}$$

Therefore, to analyze the normal curvature function, it suffices to restrict our attention to the function

$$C : S(T_{\phi(\mathbf{x})}G) \to \mathbb{R}, \tag{4.11}$$

where $S(T_{\phi(\mathbf{x})}G)$ denotes the collection of vectors tangent to the graph G with unit length.

To understand the curvature function C above, we focus at a point $\mathbf{x}_0$, which is a critical point for the function f. We do this because as Section 3.1 notes, after a rigid motion of the graph, the geometry of the graph at $\mathbf{x}_0$ is the geometry of a critical point of a graph. In addition, this is the only case that is relevant to our discussion in Part III.

Given any point $\mathbf{x}_0$, the space of vectors tangents to the graph G at the given point is spanned by $d\phi(\mathbf{e}_1), \ldots, d\phi(\mathbf{e}_n)$. Assume that our function f is regular in the sense that the space $T_{\phi(\mathbf{x})}G$ is n-dimensional, then the set $S(T_{\phi(\mathbf{x}_0)}G)$ is identified to an $(n-1)$-dimensional sphere when $T_{\phi(\mathbf{x}_0)}G$ is identified to $\mathbb{R}^n$. Assume that $\mathbf{x}_0$ is a critical point of f, according to (4.11) on this sphere,

$$C_{\mathbf{v}} = \mathbf{v}^T \ddot{\mathbf{F}}(\mathbf{x}_0)\mathbf{v}. \tag{4.12}$$

Considering $\ddot{\mathbf{F}}$, the evaluation of the Hessian at the point $\mathbf{x}_0$ is an $n \times n$-symmetric matrix. According to Proposition 1.2.3, the curvature function C has n-many critical points $\mathbf{v}_1, \ldots, \mathbf{v}_n$. There are real numbers $\lambda_1, \ldots, \lambda_n$ such that

$$|\mathbf{v}_j| = 1, \quad \langle \mathbf{v}_j, \mathbf{v}_k \rangle = 0 \quad \text{if } j \neq k, \quad \ddot{\mathbf{F}}\mathbf{v}_j = \lambda_j \mathbf{v}_j.$$

Moreover, the critical values, i.e., the values of the normal curvatures at its critical points are precisely the eigenvalues because

$$C_{\mathbf{v}_j} = \mathbf{v}_j^T \ddot{\mathbf{F}}\mathbf{v}_j = \lambda_j \mathbf{v}_j^T \mathbf{v}_j = \lambda_j \langle \mathbf{v}_j, \mathbf{v}_j \rangle = \lambda_j.$$

In a geometric context, the eigenvalues λ_j are called the *principlal curvatures* of the graph at the point $\mathbf{x}_0$. The one-dimensional space spanned by the eigenvectors $\mathbf{v}_j$ are called the *principlal directions* of the graph at the point $\mathbf{x}_0$. In future, we will choose an order for the eigenvalues, such that

$$\lambda_1 \geqslant \lambda_2 \geqslant \cdots \geqslant \lambda_{n-1} \geqslant \lambda_n. \tag{4.13}$$

We will also use $\lambda_{\max}$ for the largest eigenvalue λ_1 and $\lambda_{\min}$ for the smallest λ_n. We denote the diagonal matrix of eigenvalues as $\mathbf{\Lambda}$.

The *mean curvature* $\mu(\mathbf{x}_0)$ of the graph of a function $f : U \subseteq \mathbb{R}^n \to \mathbb{R}$ at a point $\mathbf{x}_0$ is the average of all principle curvatures

$$\overline{\lambda} = \frac{1}{n}(\lambda_1 + \cdots + \lambda_n) \tag{4.14}$$

of the trace of Hessian matrix $\ddot{\mathbf{F}}$ evaluated at $\mathbf{x}_0$:

$$\mu(\mathbf{x}_0) = \frac{1}{n} \operatorname{Tr} \ddot{\mathbf{F}}(\mathbf{x}_0).$$

Similarly, the *Guassian curvature* $\chi(\mathbf{x}_0)$ is the determinant of the Hessian matrix at $\mathbf{x}_0$. As $\det \ddot{\mathbf{F}} \det \mathbf{A} = \det \mathbf{A} \det \mathbf{\Lambda}$, the Guassian curvature is the product of the principle curvatures

$$\chi(\mathbf{x}_0) = \det \ddot{\mathbf{F}}(\mathbf{x}_0) = \lambda_1 \cdots \lambda_n. \tag{4.15}$$

As noted in Section 3.2, given a non-zero vector $\mathbf{v}$ at a point $\mathbf{x}_0$, one may consider the cross section $G(\mathbf{v})$ as a curve. In the case that $\mathbf{x}_0$ is a critical point of the function f, the graph is given by (3.11):

$$\phi(\mathbf{x}_0 + t\mathbf{v}) = \phi(\mathbf{x}_0) + td\phi(\mathbf{v}) + \nu(t)\mathbf{n},$$

where $\nu(t) = f(\mathbf{x}_0 + t\mathbf{v}) - f(\mathbf{x}_0)$. Consider it a curve in the plane spanned by $d\phi(\mathbf{v})$ and $\mathbf{n}$, then we have the map

$$t \mapsto \begin{pmatrix} t \\ \nu(t) \end{pmatrix}.$$

This is the graph of the function $\nu(t)$. Note that $\nu(0) = 0$ by construction, and $\nu'(0) = 0$ by the assumption that $\mathbf{x}_0$ is a critical point of f. To describe the concavity of the function ν in the normal plane $N(\mathbf{v})$, we calculate its second derivative.

$$\begin{aligned} \frac{d\nu}{dt} &= \sum_{j=1}^{n} \frac{\partial f}{\partial x_j}(\mathbf{x}_0 + t\mathbf{v}) v_j; \\ \frac{d^2\nu}{dt^2} &= \sum_{j,k=1}^{n} \frac{\partial^2 f}{\partial x_k \partial x_j} v_k v_j = \mathbf{v}^T \ddot{\mathbf{F}}(\mathbf{x}_0 + t\mathbf{v})\mathbf{v}. \end{aligned}$$

Therefore, when $|\mathbf{v}| = 1$, at $\mathbf{x}_0$

$$C_{\mathbf{v}} = \mathbf{v}^T \ddot{\mathbf{F}}(\mathbf{x}_0)\mathbf{v} = \frac{d^2\nu}{dt^2}_{|t=0}. \tag{4.16}$$

In other words, the normal curvature in the direction of $\mathbf{v}$ at the critical point of the function f is precisely the concavity of the normal section $G(\mathbf{v})$.

Furthermore, if the normal curvature at the point $\mathbf{x}_0$ is a constant λ, for any vector $\ddot{\mathbf{F}}(\mathbf{x}_0)\mathbf{v} = \lambda\mathbf{v}$ for all vectors. According to Taylor expansion, up to order 2, the function $f(\mathbf{x}_0 + \mathbf{v})$ is approximated by

$$f(\mathbf{x}_0) + \frac{1}{2}\mathbf{v}^T \ddot{\mathbf{F}}(\mathbf{x}_0)\mathbf{v} = f(\mathbf{x}_0) + \frac{\lambda}{2}\mathbf{v}^T\mathbf{v}.$$

In other words, near the point $\mathbf{x}_0$ and up to order 2, the function $f(\mathbf{x})$ is approximated by

$$f(\mathbf{x}_0) + \frac{\lambda}{2}(\mathbf{x} - \mathbf{x}_0)^T(\mathbf{x} - \mathbf{x}_0). \tag{4.17}$$

The graph of this function is a paraboloid.

4.2 Sectional curvatures

Suppose a function f has two variables (x_1, x_2), the graph of f is in $\mathbb{R}^3$. A point in $\mathbb{R}^3$ is on the graph if $x_3 = f(x_1, x_2)$. Suppose $f(0,0) = 0$ and it has $(0,0)$ as a minimum, then the positive x_3-axis points in the upward normal direction and the tangent plane is the x_1x_2-plane. If one applies the concept of normal curvatures to study the geometry of this graph at $\mathbf{0}$, one surveys the landscape around the origin $\mathbf{0}$ by checking the concavity in each tangent direction. However, the landscape is two-dimensional. If one wants to use a number as an indicator in the landscape there are several ways: the maximum concavity in all directions, the minimum concavity, the average of the last two numbers, the product of the last two numbers, or the average concavity in all directions. As previously noted, the maximum and minimum concavity take place in the principal directions. The numerical measures are the eigenvalues $\lambda_{\max}$ and $\lambda_{\min}$ of the Hessian $\ddot{\mathbf{F}}$ evaluated at $\mathbf{0}$. A common invariant of a matrix is its determinant. For a 2×2-matrix, it is equal to the product $\lambda_{\max}\lambda_{\min}$.

In general, let $\mathbf{v}$ and $\mathbf{w}$ be any tangent vectors of the influence graph at $\mathbf{x}_0$. Assume that they are orthonormal with respect to the first fundamental form. The three-dimensional span of $\mathbf{v}, \mathbf{w}$ and the normal vector $\mathbf{n}$ cuts across the graph along a two-dimensional surface in $\mathbb{R}^3$. Recall that the cross section of graph G in this three-dimensional space is denoted by $G(\mathbf{v}, \mathbf{w})$.

The *sectional curvature* $\mathrm{Sec}(\mathbf{v}, \mathbf{w})$ of the cross section $G(\mathbf{v}, \mathbf{w})$ at the point $\mathbf{x}_0$ is

$$\mathrm{Sec}(\mathbf{v}, \mathbf{w}) = \frac{\langle \mathbf{v}, \ddot{\mathbf{F}}\mathbf{v}\rangle\langle \mathbf{w}, \ddot{\mathbf{F}}\mathbf{w}\rangle - \langle \mathbf{v}, \ddot{\mathbf{F}}\mathbf{w}\rangle\langle \mathbf{w}, \ddot{\mathbf{F}}\mathbf{v}\rangle}{\langle \mathbf{v}, \mathbf{v}\rangle\langle \mathbf{w}, \mathbf{w}\rangle - \langle \mathbf{v}, \mathbf{w}\rangle^2}. \tag{4.18}$$

This is an elementary exercise to show that if $\{\mathbf{v}', \mathbf{w}'\}$ and $\{\mathbf{v}, \mathbf{w}\}$ spanned the same tangent subspace in $T_{\phi(\mathbf{x}_0)}S$, then $\mathrm{Sec}(\mathbf{v}, \mathbf{w}) = \mathrm{Sec}(\mathbf{v}', \mathbf{w}')$. In other words, if $G(\mathbf{v}, \mathbf{w}) = G(\mathbf{v}', \mathbf{w}')$, then $\mathrm{Sec}(\mathbf{v}, \mathbf{w}) = \mathrm{Sec}(\mathbf{v}', \mathbf{w}')$. This means that sectional curvature is a quantity concerning the plane spanned by the vectors $\mathbf{v}$ and $\mathbf{w}$, but it does not depend on the choice of any particular pair of vectors spanning the two-dimensional plane.

Exercise 4.2.1 If $\mathbf{v}' = a_1\mathbf{v} + a_2\mathbf{w}$, and $\mathbf{w}' = b_1\mathbf{w} + b_2\mathbf{w}$ and $a_1b_2 - a_2b_1 \neq 0$, then $\mathrm{Sec}(\mathbf{v}, \mathbf{w}) = \mathrm{Sec}(\mathbf{v}', \mathbf{w}')$.

If the two vectors in question are $d\phi(\mathbf{e}_i)$ and $d\phi(\mathbf{e}_j)$ in the standard basis, then the cross section is the graph of function f with respect to the variables x_i and x_j.

$$\begin{aligned}&\langle d\phi(\mathbf{e}_i), \ddot{\mathbf{F}} d\phi(\mathbf{e}_i)\rangle\langle d\phi(\mathbf{e}_j), \ddot{\mathbf{F}} d\phi(\mathbf{e}_j)\rangle \\ &-\langle d\phi(\mathbf{e}_i), \ddot{\mathbf{F}} d\phi(\mathbf{e}_j)\rangle\langle d\phi(\mathbf{e}_j), \ddot{\mathbf{F}} d\phi(\mathbf{e}_i)\rangle \\ &= \det\begin{pmatrix} \dfrac{\partial^2 f}{\partial x_i^2} & \dfrac{\partial^2 f}{\partial x_i \partial x_j} \\ \dfrac{\partial^2 f}{\partial x_i \partial x_j} & \dfrac{\partial^2 f}{\partial x_j^2} \end{pmatrix},\end{aligned} \tag{4.19}$$

$$\begin{aligned}&\langle d\phi(\mathbf{e}_i), d\phi(\mathbf{e}_i)\rangle\langle d\phi(\mathbf{e}_j), d\phi(\mathbf{e}_j)\rangle - \langle d\phi(\mathbf{e}_i), d\phi(\mathbf{e}_j)\rangle^2 \\ &= \left(1+\left(\frac{\partial f}{\partial x_i}\right)^2\right)\left(1+\left(\frac{\partial f}{\partial x_j}\right)^2\right) - \left(\frac{\partial f}{\partial x_i}\frac{\partial f}{\partial x_j}\right)^2 \\ &= 1+\left(\frac{\partial f}{\partial x_i}\right)^2+\left(\frac{\partial f}{\partial x_j}\right)^2.\end{aligned} \tag{4.20}$$

It follows that

$$\operatorname{Sec}(\mathbf{e}_i, \mathbf{e}_j) = \frac{\dfrac{\partial^2 f}{\partial x_i^2}\dfrac{\partial^2 f}{\partial x_j^2} - \left(\dfrac{\partial^2 f}{\partial x_i \partial x_j}\right)^2}{1+\left(\dfrac{\partial f}{\partial x_i}\right)^2+\left(\dfrac{\partial f}{\partial x_j}\right)^2}. \tag{4.21}$$

Proposition 4.2.2 *Suppose that $\mathbf{x}_0$ is a critical point of the function f. The sectional curvature of the graph of f at the point $\phi(\mathbf{x}_0)$ of any pair of vectors in the standard basis $\mathbf{e}_i$ and $\mathbf{e}_j$ is thus equal to*

$$\operatorname{Sec}(\mathbf{e}_i, \mathbf{e}_j) = \frac{\partial^2 f}{\partial x_i^2}\frac{\partial^2 f}{\partial x_j^2} - \left(\frac{\partial^2 f}{\partial x_i \partial x_j}\right)^2.$$

In the future, we will use $\ddot{\mathbf{F}}_{(ij)}$ to represent the 2×2-matrix:

$$\ddot{\mathbf{F}}_{(ij)} = \begin{pmatrix} \dfrac{\partial^2 f}{\partial x_i^2} & \dfrac{\partial^2 f}{\partial x_i \partial x_j} \\ \dfrac{\partial^2 f}{\partial x_i \partial x_j} & \dfrac{\partial^2 f}{\partial x_j^2} \end{pmatrix}. \tag{4.22}$$

In particular, $\operatorname{Sec}(\mathbf{e}_i, \mathbf{e}_j) = \det \ddot{\mathbf{F}}_{(ij)}$ at a critical point of the graph.

Chapter 5

Transformations and Invariance

In the last chapter, we reviewed many ways to construct quantitative measures of the concavity or curvature of the graph of a function $f : \mathcal{U} \subseteq \mathbb{R}^n \to \mathbb{R}$. In application, a change of units in the domain might change the quantitative behavior of the function f, but might not change its qualitative meaning. For example, when the unit in $\mathcal{U}$ is changed from meters to centimeters or from pounds to kilograms, the variables in $\mathcal{U}$ change. Therefore, it is necessary to investigate how a change of variables affects the various curvature measures.

Throughout this chapter, $\mathcal{U}$ is a connected open subset of $\mathbb{R}^n$ with coordinates $\mathbf{x} = (x_1, \ldots, x_n)^T$. To concentrate our effort on developing concepts useful in our application, we focus on the geometry of the graph of f at a critical point, say $\mathbf{x}_0$. In particular, its tangent space $T_{\phi(\mathbf{x}_0)}G$ is the standard embedding of $\mathbb{R}^n$ in $\mathbb{R}^{n+1}$.

5.1 Change of coordinates

A change of coordinates for functions from $\mathcal{U}$ to $\mathbb{R}$ consists of an open subset $\mathcal{V}$ in $\mathbb{R}^n$ and a differentiable map $\gamma : \mathcal{V} \to \mathcal{U}$, such that γ is a one-to-one, onto map, and at every point $\mathbf{t} = (t_1, \ldots, t_n)^T$ the differential $d\gamma$ is a non-degenerate matrix. An equivalent reformulation of the constraint is to require that the map $\gamma : \mathcal{V} \to \mathcal{U}$ have a differentiable inverse $\gamma^{-1} : \mathcal{U} \to \mathcal{V}$.

Let $\mathbf{x} = \gamma(\mathbf{t})$. The differential $d\gamma$ maps the tangent space of $\mathbb{R}^n$ to the tangent space of $\mathbb{R}^n$:

$$d\gamma : T_{\mathbf{t}}\mathbb{R}^n \to T_{\mathbf{x}}\mathbb{R}^n.$$

Treating $d\gamma(\mathbf{v})$ and $d\gamma(\mathbf{w})$ as vectors in $T_{\mathbf{x}}\mathbb{R}^n$, we compare their inner products with the inner product of $\mathbf{v}$ and $\mathbf{w}$ in $T_{\mathbf{t}}\mathbb{R}^n$.

A change of variables $\gamma : \mathcal{V} \to \mathcal{U}$ is an *isometry at a point* $\mathbf{t}$ if the differential of the map γ at the point $\mathbf{t}$, $d\gamma(\mathbf{t})$, is an orthogonal matrix. If $\gamma(\mathbf{t})$ is an orthogonal transformation for each $\mathbf{t}$ in $\mathcal{V}$, then γ is an isometry from $\mathcal{V}$ to $\mathcal{U}$.

It is a *conformal transformation at a point* $\mathbf{t}$ with factor $c > 0$ if $d\gamma(\mathbf{t})^T d\gamma(\mathbf{t}) = c^2\mathbf{I}$.

On the tangent space level, for each given point the two concepts above can be treated as identical to those introduced in Sections 2.1 and 2.4.

Example 5.1.1 Recall Theorem 2.2.2, that every rigid motion on $\mathbb{R}^n$ has the form $\gamma(\mathbf{t}) = \mathbf{At} + \mathbf{p}$ where $\mathbf{A}$ is an $n \times n$-orthogonal matrix and $\mathbf{p}$ is a vector in $\mathbb{R}^n$. As $d\gamma = \mathbf{A}$, rigid motions are examples of isometries.

Example 5.1.2 Let $\mathbf{H}$ be an $n \times n$-symmetric matrix, $\mathbf{A}$ an $n \times n$-orthogonal matrix and $\mathbf{p}$ a vector in $\mathbb{R}^n$. Define

$$\gamma(\mathbf{t}) = \mathbf{p} + \mathbf{At} + \frac{1}{2}\mathbf{t}^T\mathbf{Ht}.$$

Then $d\gamma = \mathbf{A} + \mathbf{Ht}$, and $d\gamma(\mathbf{0}) = \mathbf{A}$. Therefore, the map γ is an isometry at the point $\mathbf{0}$, although it is not necessarily an isometry at other points. Here one sees that the concept of isometries is a non-linear generalization of the concept of rigid motions.

Example 5.1.3 Let $x = \tau(t)$ be any differentiable function of a single variable, such that $\tau(0) = 0$ and $\tau'(0) \neq 0$. Define $x_j = \tau(t_j)$ for $1 \leqslant j \leqslant n$. These maps define γ as a change of coordinates from an open set containing $\mathbf{0}$ in $\mathbb{R}^n$ to another open set containing $\mathbf{0}$ in $\mathbb{R}^n$ with the property that $\gamma(\mathbf{0}) = \mathbf{0}$ and

$$d\gamma(\mathbf{0}) = \tau'(0)\mathbf{I}.$$

This is a conformal matrix and the example produces an abundant collection of non-linear conformation transformations at a point.

5.2 Non-linear conformal transformations

We now analyze the effects of non-linear conformational transformations on curvature. Let $f : \mathcal{U} \to \mathbb{R}$ be a differentiable function with $\mathbf{x}_0$ as a critical point and $\gamma : \mathcal{V} \to \mathcal{U}$ be a change of variables. The composition $g(\mathbf{t}) = f(\gamma(\mathbf{t}))$ is a differentiable function of variable $\mathbf{t}$. According to Chain Rule,

$$\frac{\partial g}{\partial t_l} = \sum_{i=1}^{n} \frac{\partial f}{\partial x_i}\frac{\partial \gamma_i}{\partial t_l}, \tag{5.1}$$

where γ_i is the i-th coordinate of the function $\gamma(\mathbf{t})$. In matrix form, it is

$$
\begin{aligned}
dg &= \left(\frac{\partial g}{\partial t_1}, \ldots, \frac{\partial g}{\partial t_n}\right) = \left(\frac{\partial f}{\partial x_1}, \ldots, \frac{\partial f}{\partial x_n}\right) \begin{pmatrix} \frac{\partial \gamma_1}{\partial t_1} & \cdots & \frac{\partial \gamma_1}{\partial t_n} \\ \vdots & \ddots & \vdots \\ \frac{\partial \gamma_n}{\partial t_1} & \cdots & \frac{\partial \gamma_n}{\partial t_n} \end{pmatrix} \\
&= df \circ d\gamma.
\end{aligned} \tag{5.2}
$$

After a transposition, the above identity becomes

$$\nabla_g = (d\gamma)^T \nabla_f. \tag{5.3}$$

By assumption, $\nabla_f(\mathbf{x}_0) = \mathbf{0}$. The above identity implies that $\nabla_g(\mathbf{x}_0) = \mathbf{0}$.

Taking second derivatives, we have

$$\frac{\partial^2 g}{\partial t_k \partial t_l} = \sum_{i=1}^{n} \frac{\partial f}{\partial x_i} \frac{\partial^2 \gamma_i}{\partial t_k \partial t_l} + \sum_{i,j=1}^{n} \frac{\partial \gamma_j}{\partial t_k} \frac{\partial^2 f}{\partial x_j \partial x_i} \frac{\partial \gamma_i}{\partial t_l}. \tag{5.4}$$

At the critical point, it is equivalent to

$$\ddot{\mathbf{G}}(\mathbf{t}_0) = d\gamma(\mathbf{t}_0)^T \ddot{\mathbf{F}}(\mathbf{x}_0) d\gamma(\mathbf{t}_0). \tag{5.5}$$

In the rest of this chapter, we denote $d\gamma(\mathbf{t}_0)$ by $\mathbf{\Gamma}$.

We are interested in comparing the graph of the function $f : \mathcal{U} \to \mathbb{R}$ with the graph of the function $g = f \circ \gamma : \mathcal{V} \to \mathbb{R}$. When $\phi : \mathcal{U} \to \mathbb{R}^{n+1}$ is an analytic representation of the graph of f, the graph of the function g is represented by the composition

$$\varphi = \phi \circ \gamma : \mathcal{V} \to \mathcal{U} \to \mathbb{R}^{n+1}.$$

Apparently, the geometry of φ is related to the geometry of ϕ through the map γ, which makes it necessary to develop this relation explicitly. We will use the superscripts g and f to indicate a geometric object in the graphs of the respective functions at their respective points, $\mathbf{t}_0$ and $\mathbf{x}_0$. For example, II^g represents the second fundamental form for the graph of g at $\mathbf{t}_0$ and C^f represents the normal curvature of the graph of the function f at $\mathbf{x}_0$.

Recall that the first fundamental form of the graph of a function at a critical point is the usual inner product in the standard embedding. Therefore, at $\mathbf{t}_0$, $\mathbb{I}^g(\mathbf{v}, \mathbf{w}) = \langle \mathbf{v}, \mathbf{w} \rangle = \mathbf{v}^T\mathbf{w}$. The differential $\mathbf{\Gamma}$ takes tangent vectors to the graph of g at $\mathbf{t}_0$ to tangent vectors to the graph of f at $\mathbf{x}_0$. Therefore, $\mathbb{I}^f(\mathbf{\Gamma v}, \mathbf{\Gamma w}) = \langle \mathbf{\Gamma v}, \mathbf{\Gamma w} \rangle = \mathbf{v}^T \mathbf{\Gamma}^T \mathbf{\Gamma w}$. If γ is a conformal transformation at $\mathbf{t}_0$ with factor c, then $\mathbf{\Gamma}^T\mathbf{\Gamma} = c^2\mathbf{I}$ and

$$c^2 \mathbb{I}^g(\mathbf{v}, \mathbf{w}) = \mathbb{I}^f(\mathbf{\Gamma v}, \mathbf{\Gamma w}) = c^2 \mathbf{v}^T \mathbf{w}.$$

As a consequence of Proposition 4.1.1, the normal curvature of the graph of the function g at $\mathbf{t}_0$ is equal to

$$C^g_{\mathbf{v}} = \frac{\langle \ddot{\mathbf{G}}\mathbf{v}, \mathbf{v}\rangle}{\langle \mathbf{v}, \mathbf{v}\rangle} = \frac{\mathbf{v}^T \ddot{\mathbf{G}}\mathbf{v}}{\mathbf{v}^T\mathbf{v}},$$

where $\mathbf{v}$ is any tangent vector. According to (5.5), it is equal to

$$\frac{\mathbf{v}^T\mathbf{\Gamma}^T\ddot{\mathbf{F}}\mathbf{\Gamma}\mathbf{v}}{\mathbf{v}^T\mathbf{v}} = \frac{(\mathbf{\Gamma}\mathbf{v})^T\ddot{\mathbf{F}}(\mathbf{\Gamma}\mathbf{v})}{\mathbf{v}^T\mathbf{v}}.$$

On the other hand, at $\mathbf{x}_0$,

$$C^f_{\mathbf{\Gamma}\mathbf{v}} = \frac{(\mathbf{\Gamma}\mathbf{v})^T\ddot{\mathbf{F}}(\mathbf{\Gamma}\mathbf{v})}{(\mathbf{\Gamma}\mathbf{v})^T\mathbf{\Gamma}\mathbf{v}} = \frac{(\mathbf{\Gamma}\mathbf{v})^T\ddot{\mathbf{F}}(\mathbf{\Gamma}\mathbf{v})}{\mathbf{v}^T\mathbf{\Gamma}^T\mathbf{\Gamma}\mathbf{v}}.$$

When $\mathbf{\Gamma}^T\mathbf{\Gamma} = c^2\mathbf{I}$,

$$C^f_{\mathbf{\Gamma}\mathbf{v}} = \frac{1}{c^2}C^g_{\mathbf{v}}.$$

Proposition 5.2.1 *Suppose $\mathcal{U}$ is a connected open subset of $\mathbb{R}^n$, and $f : \mathcal{U} \to \mathbb{R}$ is a differentiable function with $\mathbf{x}_0$ as a critical point. If a change of variables $\gamma : \mathcal{V} \to \mathcal{U}$, such that $\gamma(\mathbf{t}_0) = \mathbf{x}_0$, and $\mathbf{\Gamma}^T\mathbf{\Gamma} = c^2\mathbf{I}$ where $\mathbf{\Gamma} = d\gamma(\mathbf{t}_0)$, then for any tangent vectors $\mathbf{v}, \mathbf{w}$ of the graph of g at $\mathbf{t}_0$,*

$$\mathbb{I}^f(\mathbf{\Gamma}\mathbf{v}, \mathbf{\Gamma}\mathbf{w}) = c^2\mathbb{I}^g(\mathbf{v}, \mathbf{w}), \quad C^g_{\mathbf{v}} = c^2 C^f_{\mathbf{\Gamma}\mathbf{v}}. \tag{5.6}$$

Similarly, by definition (4.18) the sectional curvature for the graphs of g and f are respectively:

$$\begin{aligned}\mathrm{Sec}^g(\mathbf{v}, \mathbf{w}) &= \frac{\langle \mathbf{v}, \ddot{\mathbf{G}}\mathbf{v}\rangle\langle \mathbf{w}, \ddot{\mathbf{G}}\mathbf{w}\rangle - \langle \mathbf{v}, \ddot{\mathbf{G}}\mathbf{w}\rangle\langle \mathbf{w}, \ddot{\mathbf{G}}\mathbf{v}\rangle}{\langle \mathbf{v}, \mathbf{v}\rangle\langle \mathbf{w}, \mathbf{w}\rangle - \langle \mathbf{v}, \mathbf{w}\rangle^2} \\ &= \frac{\langle \mathbf{\Gamma}\mathbf{v}, \ddot{\mathbf{F}}\mathbf{\Gamma}\mathbf{v}\rangle\langle \mathbf{\Gamma}\mathbf{w}, \ddot{\mathbf{F}}\mathbf{\Gamma}\mathbf{w}\rangle - \langle \mathbf{\Gamma}\mathbf{v}, \ddot{\mathbf{F}}\mathbf{\Gamma}\mathbf{w}\rangle\langle \mathbf{\Gamma}\mathbf{w}, \ddot{\mathbf{F}}\mathbf{\Gamma}\mathbf{v}\rangle}{\langle \mathbf{v}, \mathbf{v}\rangle\langle \mathbf{w}, \mathbf{w}\rangle - \langle \mathbf{v}, \mathbf{w}\rangle^2},\end{aligned}$$

$$\mathrm{Sec}^f(\mathbf{\Gamma}\mathbf{v}, \mathbf{\Gamma}\mathbf{w}) = \frac{\langle \mathbf{\Gamma}\mathbf{v}, \ddot{\mathbf{F}}\mathbf{\Gamma}\mathbf{v}\rangle\langle \mathbf{\Gamma}\mathbf{w}, \ddot{\mathbf{F}}\mathbf{\Gamma}\mathbf{w}\rangle - \langle \mathbf{\Gamma}\mathbf{v}, \ddot{\mathbf{F}}\mathbf{\Gamma}\mathbf{w}\rangle\langle \mathbf{\Gamma}\mathbf{w}, \ddot{\mathbf{F}}\mathbf{\Gamma}\mathbf{v}\rangle}{\langle \mathbf{\Gamma}\mathbf{v}, \mathbf{\Gamma}\mathbf{v}\rangle\langle \mathbf{\Gamma}\mathbf{w}, \mathbf{\Gamma}\mathbf{w}\rangle - \langle \mathbf{\Gamma}\mathbf{v}, \mathbf{\Gamma}\mathbf{w}\rangle^2}.$$

It follows that when $\mathbf{\Gamma}^T\mathbf{\Gamma} = c^2\mathbf{I}$,

$$\mathrm{Sec}^f(\mathbf{\Gamma}\mathbf{v}, \mathbf{\Gamma}\mathbf{w}) = \frac{1}{c^4}\mathrm{Sec}^g(\mathbf{v}, \mathbf{w}).$$

5.3 Invariant curvatures

Recall equation (5.5) that at the critical point $\mathbf{t}_0$, $\ddot{\mathbf{G}} = \mathbf{\Gamma}^T\ddot{\mathbf{F}}\mathbf{\Gamma}$. Because $\mathbf{\Gamma}^T\mathbf{\Gamma} = c^2\mathbf{I}$, $\mathbf{\Gamma}^T = c^2\mathbf{\Gamma}^{-1}$. Therefore, the mean curvatures of the graphs of functions g

and f over the points $\mathbf{t}_0$ and $\mathbf{x}_0$, respectively, are related by the identity

$$\mathrm{Tr}\,\ddot{\mathbf{G}} = c^2 \mathrm{Tr}\left(\mathbf{\Gamma}^{-1}\ddot{\mathbf{F}}\mathbf{\Gamma}\right) = c^2 \mathrm{Tr}\,\ddot{\mathbf{F}}. \tag{5.7}$$

As long as it does not equal to zero, we can consider the quantities

$$A^g_{\mathbf{v}} = \frac{1}{\mathrm{Tr}\,\ddot{\mathbf{G}}} C^g_{\mathbf{v}} = \frac{1}{\mathrm{Tr}\,\ddot{\mathbf{G}}} \frac{\mathbf{v}^T\ddot{\mathbf{G}}\mathbf{v}}{\mathbf{v}^T\mathbf{v}}, \tag{5.8}$$

$$A^f_{\boldsymbol{\ell}} = \frac{1}{\mathrm{Tr}\,\ddot{\mathbf{F}}} C^g_{\boldsymbol{\ell}} = \frac{1}{\mathrm{Tr}\,\ddot{\mathbf{F}}} \frac{\boldsymbol{\ell}^T\ddot{\mathbf{F}}\boldsymbol{\ell}}{\boldsymbol{\ell}^T\boldsymbol{\ell}}. \tag{5.9}$$

Based on (5.6) and (5.7) above, we find that

$$A^g_{\mathbf{v}} = A^f_{\mathbf{\Gamma v}}. \tag{5.10}$$

In other words, the ratio between the normal curvature and the mean curvature at a critical point is independent of a conformal transformation at a critical point. We consider this ratio to be one version of conformal invariant curvature.

However, the above computation is valid only if the mean curvatures of the graphs at the points in question are not equal to zero. As the mean curvature is the average values of the eigenvalues of the Hessian matrices, it could be equal to zero without having a Hessian matrix equal to zero. Therefore, we consider a slightly different form of conformal invariant curvature.

Since $\ddot{\mathbf{F}}$ is a symmetric $n \times n$ matrix, there exists an orthonormal basis of eigenvectors $\{\mathbf{v}_1, \ldots, \mathbf{v}_n\}$. Let $\lambda_1, \ldots, \lambda_n$ be the corresponding eigenvalues. Consider the matrix $\mathbf{A}$ whose columns are the chosen eigenvectors:

$$\mathbf{A} = (\mathbf{v}_1, \mathbf{v}_2, \ldots, \mathbf{v}_n),$$

and define $\mathbf{\Lambda} = \mathrm{Diag}(\lambda_1, \lambda_2, \ldots, \lambda_n)$ as the diagonal matrix of eigenvalues, then $\ddot{\mathbf{F}}\mathbf{A} = \mathbf{A}\mathbf{\Lambda}$. Because the columns of $\mathbf{A}$ are orthonormal vectors, the matrix A is invertible. In fact, $\mathbf{A}^{-1} = \mathbf{A}^T$. It follows that $\ddot{\mathbf{F}}^2 = \mathbf{A}\mathbf{\Lambda}^2\mathbf{A}^{-1}$, and

$$\mathrm{Tr}\,\ddot{\mathbf{F}}^2 = \mathrm{Tr}\,\mathbf{\Lambda}^2 = \lambda_1^2 + \cdots + \lambda_n^2. \tag{5.11}$$

Therefore, this quantity is equal to zero if and only if all eigenvalues are equal to zero. It is equivalent to the matrix $\ddot{\mathbf{F}}$ being identical zero at the point $\mathbf{x}_0$. Since $\ddot{\mathbf{G}} = \mathbf{\Gamma}^T\ddot{\mathbf{F}}\mathbf{\Gamma}$, $\ddot{\mathbf{G}}$ and $\ddot{\mathbf{F}}$ are symmetric matrices,

$$\begin{aligned}
&\mathrm{Tr}\,\ddot{\mathbf{G}}^2 \\
&= \mathrm{Tr}\,(\ddot{\mathbf{G}}^T\ddot{\mathbf{G}}) = \mathrm{Tr}\,(\mathbf{\Gamma}^T\ddot{\mathbf{F}}^T\mathbf{\Gamma}\mathbf{\Gamma}^T\ddot{\mathbf{F}}\mathbf{\Gamma}) = \mathrm{Tr}\,(c^2\mathbf{\Gamma}^{-1}\ddot{\mathbf{F}}^T\ddot{\mathbf{F}}\mathbf{\Gamma}) \\
&= c^4\mathrm{Tr}\,(\ddot{\mathbf{F}}^T\ddot{\mathbf{F}}) = c^4\mathrm{Tr}\,\ddot{\mathbf{F}}^2.
\end{aligned}$$

Therefore,

$$\sqrt{\operatorname{Tr}\ddot{\mathbf{G}}^2} = c^2\sqrt{\operatorname{Tr}\ddot{\mathbf{F}}^2}. \tag{5.12}$$

Consider the quantities

$$B^g_{\mathbf{v}} = \frac{1}{\sqrt{\operatorname{Tr}\ddot{\mathbf{G}}^2}} C^g_{\mathbf{v}} = \frac{1}{\sqrt{\operatorname{Tr}\ddot{\mathbf{G}}^2}} \frac{\mathbf{v}^T\ddot{\mathbf{G}}\mathbf{v}}{\mathbf{v}^T\mathbf{v}}, \tag{5.13}$$

$$B^f_{\boldsymbol{\ell}} = \frac{1}{\sqrt{\operatorname{Tr}\ddot{\mathbf{F}}^2}} C^f_{\boldsymbol{\ell}} = \frac{1}{\sqrt{\operatorname{Tr}\ddot{\mathbf{F}}^2}} \frac{\boldsymbol{\ell}^T\ddot{\mathbf{F}}\boldsymbol{\ell}}{\boldsymbol{\ell}^T\boldsymbol{\ell}}. \tag{5.14}$$

According to (5.6) and (5.12), we find that

$$B^g_{\mathbf{v}} = B^f_{\mathbf{\Gamma v}}. \tag{5.15}$$

This quantity is another version of conformal invariant curvature at the critical point of the graph of a function.

Because the ratio of $A^f_{\boldsymbol{\ell}}$ to $B^f_{\boldsymbol{\ell}}$ is the number

$$\frac{\operatorname{Tr}\ddot{\mathbf{F}}}{\sqrt{\operatorname{Tr}\ddot{\mathbf{F}}^2}},$$

which is independent of $\boldsymbol{\ell}$, future applications will focus on $B_{\boldsymbol{\ell}}$, given its generality.

Part II

Statistics

Chapter 6

Discrete Random Variables and Related Concepts

In Part II, we introduce some basic statistics concepts that help the content of this book remained self-contained and are the most relevant to our subsequent discussion in Part III.

6.1 Preliminaries

Statistics is a science that deals with data, which are collected using various methods. Statistical analysis is frequently based on a data set that serves as a *random sample* from the target *population* . Simply speaking, a population is the collection of the measurements for one or more characteristics from a group of subjects or objects that are of interest in a study. A random sample is randomly a selected subset of the population. For example, if we want to study the average height of male undergraduate students in a university, then a collection of the heights of all of male undergraduate students constitutes the target population. If we randomly draw a group of male students from this target population and measure their heights, we obtain a random sample.

Based on the information regarding extracted from such as sample of students' heights, we can draw inferences regarding the height of the entire population of male undergraduate students. For example, we may use the average height of the sample (*sample mean*) to estimate the average height of the entire population (*population mean*). The techniques used to draw inferences regarding the population through sample information are known as *statistical inference* techniques. To facilitate the drawing of statistical inferences, we may need to impose sensible assumptions on the population. For example, to use the sample data to infer the average height of the population of male undergraduate students, it is convenient if we assume that the distribution of the heights of the entire population of male undergraduate students follows a *normal distribution*. A normal distribution is a typical distribution for a *continuous random variable*, which will be introduced in Chapter 7.

If we want to study two or more characteristics, such as both the weight and the height of male undergraduate students, we must consider two or more variables simultaneously in a bivariate or multivariate analysis. This will be addressed in Chapter 8. To facilitate the study of the relationship between two variables, we sometimes assume that the relationship is linear and follows a *linear regression model*, which will be discussed in Chapters 9 and 10.

The set of assumptions imposed on a population, which is typically represented by a set of mathematical equations and notations, constitutes a *statistical model*. An important component in a statistical model is the *random variable*. We introduce *discrete random variables* and their relevant concepts in this chapter and address *continuous random variables* and their relevant concepts in Chapter 7.

6.2 Discrete random variables

6.2.1 Discrete random variables and probability function

In simple terms a discrete variable assumes the value in a set $\mathcal{R}$ with a finite (or countably infinite) number of possible values; if the occurrence of the value is associated with a probability, then the variable is called a **discrete random variable**. A capital letter is typically used to denote a random variable and the corresponding lower case letter is used to denote the value of the random variable.

Example 6.2.1 (Discrete random variables: Household size) Suppose it is known that the household sizes in a town with 100,000 households are given in Table 6.1.

Table 6.1 Household size

Household size	1	2	3	4	Total
Number of households	10,000	20,000	40,000	30,000	100,000

If we randomly select a household in this town and let X be the size of the household selected, then X is a discrete random variable. It can take on any one of the four possible values, namely $x = 1$, 2, 3 or 4, and we know that the probability of observing a household size of 3 is different from that of observing a household size of 4. Based on Table 6.1, these probabilities are, respectively,

0.4 for $x = 3$ and 0.3 for $x = 4$.

A *probability function* $f(x) = \Pr(X = x)$ specifies the probability of occurrence for the possible value x of a discrete random variable. The probability is a function of x.

The probability function for a discrete random variable is also known as a *frequency function*, a *probability mass function* or a *probability density function* with the term "density" more commonly used for probability functions that are associated with a *continuous random variable*, as introduced in Chapter 7.

A probability function for a discrete random variable has the following properties:

- $0 < f(x) \leqslant 1 \quad \text{for all } x \in \mathcal{R}, \text{and}$ (6.1)
- $\sum_{x \in \mathcal{R}} f(x) = 1.$ (6.2)

The possible values of a discrete random variable X, together with the associated probability function $f(x)$, specifically the set of pairs $\{(x, f(x))\}$, constitutes the *probability distribution* of X. Using Example 6.2.1, the probability distribution for the random variable $X =$ *Household size* is given in Table 6.2.

Table 6.2 The probability distribution of $X =$ Household size

Household size x	1	2	3	4
Probability $f(x)$	0.1	0.2	0.4	0.3

We use an artificial example with a known population here to better illustrate the concepts. In practical analysis, the probability distribution of the population is usually unknown and one must to draw a sample from the population for an analysis.

If we draw a sample of n households from the town and use X_i to denote the size of the household of the i-th draw, then each of the $X_i, i = 1, \ldots, n$ is a random variable. If the sample is drawn *with replacement* in the sense that a selected household is put back before the next household is drawn such that it can be drawn again, then the values that each of the X_i's can take on, together with the corresponding probabilities, are all the same as given in Table 6.2. In this case, the X_i's are considered to be *identically distributed.* Moreover, what has been observed for X_i provides no information on what would probably be observed for X_j when $i \neq j$. The distribution of X_j does not change according to the additional information provided by the observed value of X_i. This phenomenon is that X_i is *independent* of X_j. A formal definition of independence will be given in Section 8.6. The set of *identically independently distributed* (usually abbreviated as *i.i.d*) random variables $X_i, i = 1, \ldots, n$ constitutes a *random sample* of X or a random sample of the population that is represented by the random variable X.

A random sample of households can also be selected by *sampling without replacement*, in which any selected household cannot be drawn again. In such a case, the distributions of the X_i's are not identical. The distribution of X_1 is given in Table 6.2, but the distribution of X_2 is no longer the one given in Table 6.2. For example, if it is observed that $X_1 = 2$, then the probability of $X_2 = 2$ is reduced to 19999/99999. In this case, the observed value of X_1 provides extra information on what would probably be observed for X_2, and hence the two variables are not independent.

Although the differences between sampling with and without replacement should be noted, especially when the population size N is finite and small, many practical applications often neglect this difference. In many applications, the non-replacement effect is negligible when the sample size n is small relative to the population size N and the random variables $X_i (i = 1, \ldots, n)$ can still be regarded as *i.i.d* in practice.

Example 6.2.2 (Discrete random variables: Flip a coin twice) We have a coin with two sides H (head) and T (tail), and we are interested in the number of times the coin will land heads up in two flips. If the coin is fair, then the two sides are equally likely to occur. As a result, each of the four possible outcomes {HH, HT, TH, TT} is equally likely to occur in two flips and hence each occurs with a probability of 1/4; that is, if we repeatedly flip the coin twice and repeat N many times, each pattern occurs respectively 1/4 of the N times if N is sufficiently large. Let $X =$ *Number of heads in two flips*, then X is a random variable. Using the fact that each pattern occurs with a probability of 1/4, we obtain Table 6.3. The possible x values, together with their probabilities given by $f(x)$ provide the probability distribution of the random variable $X =$ *Number of heads in two flips*.

Table 6.3 Flip a coin twice

Number of heads x	0	1	2
Probability $f(x)$	0.25	0.5	0.25

If we flip the coin twice for n times, we have a set of n *i.i.d.* random variables $X_i (i = 1, \ldots, n)$. The collection $\{X_1, \ldots, X_n\}$ is a random sample of X or a random sample of the distribution with probability function $f(x)$.

We use the artificial *Household size* example in Example 6.2.1 with a known and finite population size N to introduce the basic concept of population and random variables, while the *Flip a coin twice* example in Example 6.2.2 illustrates the same concept in a wider context when the population is not as clearly and concretely defined as in Example 6.2.1. The interpretations of many statistical concepts in relation to random variables such as *mathematical expectation*, to be discussed in Section 6.3.1, can be easily comprehended in the context of a known and finite population and generalized in a way similar to the previously discussed for Example 6.2.2 if the population is not as clearly defined as that in Example 6.2.1.

Exercise 6.2.3 Throw a pair of fair dice and let X be the sum of the two numbers showing on the face of dices, then derive the distribution of X.

6.2.2 Relative frequency histogram

A *relative frequency histogram* is a graphical representation of the probability function $f(x)$. For example, the relative frequency histogram for the random variable X = *Number of heads in two flips* of a fair coin in Example 6.2.2 is given in Figure 6.1 as is a graphical representation of the probability function in Table 6.3.

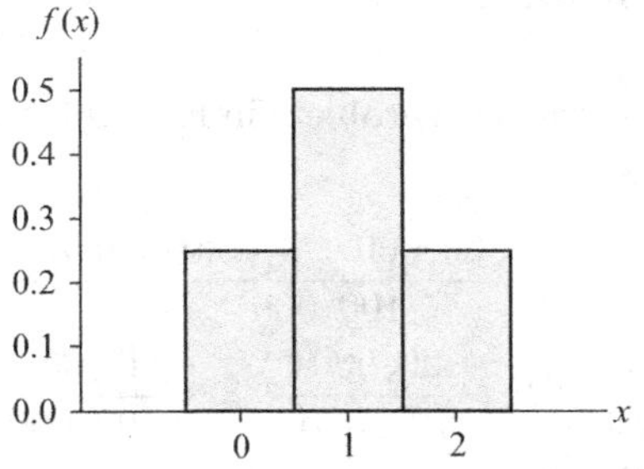

Fig. 6.1 Relative frequency histogram

The relative frequency histogram is constructed in a way such that the area of the rectangle corresponds to each x value is given by $f(x)$. Using the property that $\sum_{x \in \mathcal{R}} f(x) = 1$ (see (6.2)), the total of the areas of the rectangles equals to 1. Moreover, if the widths of the rectangles are the same, the height of each rectangle is proportional to the value of $f(x)$.

6.2.3 Cumulative distribution function

Given the probability function $f(x)$ of a random variable X, then $F(x) = \sum_{t \in \mathcal{R}, t \leqslant x} f(t) = \Pr(X \leqslant x)$ is called the *cumulative distribution function* and represents the probability that the random variable is less than or equal to a specific x value.

From Table 6.3, the cumulative distribution function of X = *Number of heads in two flips* is given in Table 6.4.

Table 6.4 Cumulative distribution function $F(x)$

Number of heads x	0	1	2
Probability function $f(x) = \Pr(X = x)$	0.25	0.5	0.25
Cumulative distribution function $F(x) = \Pr(X \leqslant x)$	0.25	0.75	1.0

Exercise 6.2.4 Throw a pair of fair dice and let X be the sum of the two numbers showing on the face of dices, then derive the cumulative distribution function of X.

6.3 Population parameters and sample statistics

6.3.1 Population mean and expected value

Continue with the *Household size* problem in Example 6.2.1. The average household size in the town is

$$\begin{aligned}
\mu &= \frac{1 \times 10,000 + 2 \times 20,000 + 3 \times 40,000 + 4 \times 30,000}{100,000} \\
&= 1 \times \frac{10,000}{100,000} + 2 \times \frac{20,000}{100,000} + 3 \times \frac{40,000}{100,000} + 4 \times \frac{30,000}{100,000} \\
&= 1 \times 0.1 + 2 \times 0.2 + 3 \times 0.4 + 4 \times 0.3 \\
&= 2.9 \\
&= \sum x f(x) \qquad (6.3)
\end{aligned}$$

where the summation is taken over all $x \in \mathcal{R}$. This shows that the *population mean* μ is equal to the sum of the products of each possible value of the random variable X and its probability of occurrence.

In general, let X be a discrete random variable with a value of $x \in \mathcal{R}$ and a probability function of $f(x)$, then

$$\mu = \mathrm{E}(X) = \sum_{x \in \mathcal{R}} x f(x) \qquad (6.4)$$

is called the *expected value* or the *mathematical expectation* of the random variable X. More details on mathematical expectation are given in Section 6.4. From (6.3), the expected value $E(X)$ is also the population mean. This interpretation of population mean can also be established in a wider context when the population is not finite or not concrete. Please see the discussion following Example 6.2.2 in Section 6.2.1.

6.3.2 Sample statistic

The expected value E(X) that is the population mean μ describes a characteristic of the population. It is a *population parameter.* Population parameters are usually unknown and are the interests of a study. Statistical analysis involves drawing inferences regarding the population parameters using sample information.

A quantity that is a function of the sample data is called a *statistic.* A statistic is a sample-level quantity. Meaningful sample statistics can extract representative sample information to characterize the sample and draw inferences regarding the population. These are called *summary statistics.* The *sample mean* and the *sample variance* are typical summary statistics, that provide information on the location and dispersion of the data set.

6.3.3 Sample mean

Let $\{X_1, \ldots, X_n\}$ be a sample that is identically and independently distributed according to the distribution $f(x)$, then $\overline{X} = \sum_{i=1}^{n} X_i$ is called the *sample mean.* It is a sample-level quantity that provides information on the central location of the sample and can be used to estimate or draw inferences regarding the population mean.

Unlike population parameters which have a fixed and usually unknown true value, a sample statistic may assume many different values because different random samples can be obtained from the same population. These different values for the sample statistic occur with different probabilities that depend on the distribution of the population from which the sample is drawn. In other words, a sample statistic is also a random variable associated with a distribution. It is common practice to use $\overline{X} = (\sum X_i)/n$ to represent the random variable of sample mean and $\bar{x} = (\sum x_i)/n$ to represent the mean value obtained from a particular sample. $\overline{X}$ as a random variable can take different $\bar{x}$ values. The distribution of a sample statistic is called a *sampling distribution* and can usually be derived if one has information on the original population. More details on sampling distribution are available in Section 7.8.

6.3.4 Sample and population variances

Sample mean $\overline{X}$ and population mean μ are basic quantities used to summarize data. They provide information on the *central tendency* of a data set and are *location measures* at the sample and population levels, respectively. In addition to central tendency, *dispersion measures* are also essential for summarizing data. A convenient measure of the dispersion of a data set is the *range*, specifically the difference between the largest and smallest values in the data set. Range is determined solely using the two most extreme data points, and hence is highly influenced by the extreme values in a data set.

Alternatively, the most typical and frequently used dispersion measure is the *variance* or, equivalently, the *standard deviation* . The variance takes the *deviation scores* of all of the data points into account. The *deviation score* of a data point is the difference between the value of the data point and the mean of the data set; that is, $x_i - \bar{x}$ for a sample data point and $x_i - \mu$ for a population data point. Deviation scores carry information on the dispersion of individual data points. However, as

$$\sum_{i=1}^{n}(x_i - \bar{x}) = 0 \quad \text{and} \quad \sum_{i=1}^{N}(x_i - \mu) = 0, \tag{6.5}$$

the average of the deviation scores cannot serve as a summary measure of the dispersion of a data set. An intuitive measure is the *mean of absolute deviations*, which is the average of the absolute values of all of the deviation scores. However, the function of absolute value is not smooth and hence hinders the development of concise mathematical and statistical theories. As a result, the average of the squares of the deviation scores is preferable and the resulting measure is the *variance* of the data set.

Population variance is a population parameter that is usually denoted by σ^2. If the population has N elements $\{x_1, \ldots, x_N\}$, then the population variance is given by

$$\sigma^2 = \frac{\sum_{i=1}^{N}(x_i - \mu)^2}{N}. \tag{6.6}$$

Figure 6.2 illustrates the meaning of the variance of a data set with $N = 4$ data points, which is the average of the areas of the squares that are determined by the deviation scores for the data points.

The positive square root of the variance, denoted by σ, is called the *standard deviation* of the data set and its unit of measurement is the same as that of the original data.

Example 6.3.1 For the household size example in Example 6.2.1, the mean of the population is $\mu = 2.9$. The population variance, according to (6.6), is given by

$$\begin{aligned}
\sigma^2 &= \frac{1}{100,000}\{(1-2.9)^2 \times 10,000 + (2-2.9)^2 \times 20,000 \\
&\quad +(3-2.9)^2 \times 40,000 + (4-2.9)^2 \times 30,000\} \\
&= (1-2.9)^2 \times \frac{10,000}{100,000} + (2-2.9)^2 \times \frac{20,000}{100,000} \\
&\quad +(3-2.9)^2 \times \frac{40,000}{100,000} + (4-2.9)^2 \times \frac{30,000}{100,000} \\
&= (1-2.9)^2 \times 0.1 + (2-2.9)^2 \times 0.2 \\
&\quad +(3-2.9)^2 \times 0.4 + (4-2.9)^2 \times 0.3 \\
&= 0.89 \\
&= \sum (x-\mu)^2 f(x). \qquad (6.7)
\end{aligned}$$

This suggests that similar to how we define population mean in Section 6.3.1, we can define the population variance using mathematical expectation as follows.

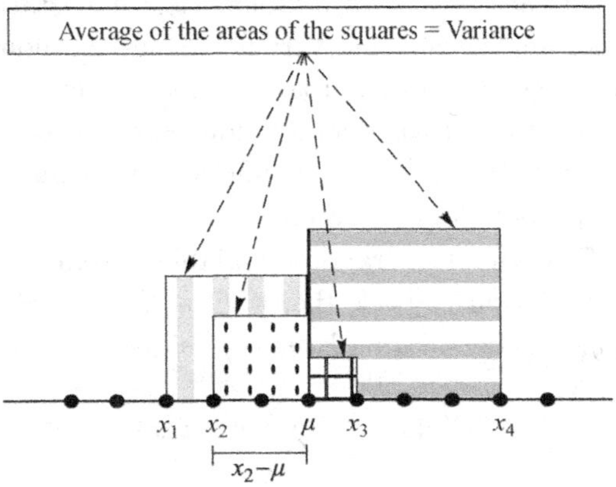

Fig. 6.2 Variance

Let X be a discrete random variable with values $x \in \mathcal{R}$ and probability function $f(x)$, such that the *variance* of a random variable X, usually denoted by σ^2 or $\mathrm{Var}(X)$, is given by

$$\sigma^2 = \mathrm{Var}(X) = \mathrm{E}(X-\mu)^2 = \sum_{x \in \mathcal{R}} (x-\mu)^2 f(x).$$

At the sample level, let $\{X_i, i = 1, \ldots, n\}$ be a random sample, then the sample variance is given by

$$S^2 = \frac{\sum_{i=1}^{n} (X_i - \overline{X})^2}{n-1}.$$

Although S^2 is meant to be the average of the squares of the deviation scores, its denominator is $n-1$ instead of n. S^2 is a sample statistic and its possible values s^2 are determined by the observed values $\{x_1, x_2, \ldots, x_n\}$. S^2 is also a random variable that is a function of the random sample $\{X_1, X_2, \ldots, X_n\}$. Because S^2 is a random variable, we can compute its expected value $\mathrm{E}(S^2)$, which can be interpreted as the mean of the population of S^2. The use of $n-1$ rather than n in the denominator ensures that $\mathrm{E}(S^2) = \sigma^2$, or that S^2 is an *unbiased estimator* of the population variance. Further discussion on unbiased estimators is provided in Section 7.8.

6.4 Mathematical expectations

We have introduced the concept of mathematical expectation in the context of population mean and population variance using the household size example (Equation (6.3) and (6.7)) to illustrate. This concept of *mathematical expectation* can be generalized to other random variable functions and can be interpreted as a population average of functions in a manner similar to the population mean and population variance.

Let $g(X)$ be a function of the random variable X that can assume the value $x \in \mathcal{R}$ and its distribution is $f(x)$, then the *mathematical expectation* or the *expected value* of $g(X)$, denoted by $\mathrm{E}(g(X))$, is given by

$$\mathrm{E}(g(X)) = \sum_{x \in \mathcal{R}} g(x) f(x). \tag{6.8}$$

Mathematical expectation possesses a number of properties and because these properties are valid for both discrete and continuous random variables, we present them in Section 7.4 after introducing the relevant concepts for continuous random variables.

Example 6.4.1 Let $g(X) = X$, we obtain the population mean $\mu = \mathrm{E}(g(X)) = \mathrm{E}(X) = \sum_{x \in \mathcal{R}} x f(x) = \mu$.

Example 6.4.2 Let $g(X) = (X - \mu)^2$, we obtain the population variance $\mathrm{Var}(X) = \mathrm{E}(g(X)) = \mathrm{E}(X - \mu)^2 = \sum_{x \in \mathcal{R}} (x - \mu)^2 f(x) = \sigma^2$.

6.5 Maximum likelihood estimation

The distribution of a random variable is represented by a probability function $f(x)$. If we know the probability density function, then we have a comprehensive picture of the random variable or the distribution. Distributions that can be represented by the same mathematical formula involving a set of unknown parameters are grouped in a family, and every set of feasible values for this set of unknown parameters determines a specific distribution. When the values for the set of parameters are known, the distribution is known.

In practical analysis, we usually have partial knowledge about the distribution of the population. With reference to this knowledge, we formulate a statistical model that includes the specifications for an appropriate family of distribution to the population. Based on a random sample that is drawn from the population, we find estimates for the unknown parameters. The *maximum likelihood method* is a typical method for finding these parameter estimates.

Let the distribution of X be $f(x) = f(x, \boldsymbol{\theta})$, where $\boldsymbol{\theta} = (\theta_1, \ldots, \theta_q)^T$ is a vector of the unknown parameters that characterize the distribution. Our objective is to find an estimate of $\boldsymbol{\theta}$ based on a random sample that is drawn from the population. Let $\{X_1, \ldots, X_n\}$ be the random sample, then each X_i is identically and independently distributed according to the probability distribution $f(x, \boldsymbol{\theta})$. Let $x_1, \ldots, x_n$ be the set of observed values for the sample, such that the probability of observing the set of values is given by

$$L^*(\boldsymbol{\theta}) = \Pr(X_1 = x_1, \ldots, X_n = x_n) = f(x_1, \boldsymbol{\theta}) f(x_2, \boldsymbol{\theta}) \cdots f(x_n, \boldsymbol{\theta}). \qquad (6.9)$$

This function $L^*(\boldsymbol{\theta})$ is a function of $\boldsymbol{\theta}$ and is called the *likelihood function.*

The values $\{x_1, x_2, \ldots, x_n\}$ are the observed sample values, which have already been realized and hence, the set of values for the elements in $\boldsymbol{\theta}$ that can produce the largest probability of the occurrence of the observed set of $\{x_1, x_2, \ldots, x_n\}$ values is more likely to be the set of true values for $\boldsymbol{\theta}$. In other words, among all possible $\boldsymbol{\theta}$ values the one that maximizes the probability or the likelihood function in (6.9) is an intuitively reasonable estimate of $\boldsymbol{\theta}$. This is the *method of maximum likelihood (ML) estimation.* The method can be applied to estimate one or more parameters.

Specifically, let $\{X_1, \ldots, X_n\}$ be a random sample from a distribution with probability function $f(x, \boldsymbol{\theta})$, where $\boldsymbol{\theta} = (\theta_1, \theta_2, \ldots, \theta_q)^T$ is a vector of the unknown parameters that characterize the distribution. The **maximum likelihood estimate** $\widehat{\boldsymbol{\theta}} = (\widehat{\theta_1}, \ldots, \widehat{\theta_q})^T$ of $\boldsymbol{\theta}$ is obtained by maximizing the *likelihood function*

$$L^*(\boldsymbol{\theta}) = \prod_{i=1}^{n} f(x_i, \boldsymbol{\theta}), \qquad (6.10)$$

with respect to $\boldsymbol{\theta}$, where $x_i, i = 1, \ldots, n$ are the observed values of $X_1, \ldots, X_n$.

In many applications, it is more convenient to maximize the log-*likelihood*, which is the natural logarithm of the likelihood function, namely $L(\boldsymbol{\theta}) = \ln L^*(\boldsymbol{\theta})$. Thus the ML estimates $\widehat{\theta}_i$ of θ_i, $i = 1, \ldots, q$ are the solutions of the *likelihood equations* :

$$\frac{\partial L(\theta)}{\partial \theta_i} = 0 \qquad \text{for} \quad i = 1, \ldots, q. \tag{6.11}$$

Each of the ML estimates $\widehat{\theta}_i = h_i(x_1, x_2, \ldots, x_n)$ of θ_i, $i = 1, \ldots, q$ are a function of the sample observed values $x_i, i = 1, \ldots, n$. They are obtained by maximizing (6.10) or solving (6.11). Expressing $\widehat{\theta}_i$ as a function of the random variables that represent the sample rather than as true observed values, the quantity $\widehat{\theta}_i = h_i(X_1, X_2, \ldots, X_n)$ is then called the *maximum likelihood estimator* of θ_i. The maximum likelihood estimators $\widehat{\theta}_i = h_i(X_1, X_2, \ldots, X_n)$, $i = 1, \ldots, q$ are functions of the sample and are sample statistics. They are random variables and each is accompanied by a probability distribution, which is also called a *sampling distribution.* The concepts on sampling distributions are discussed further in Section 7.8.

6.6 Maximum likelihood estimation of the probability of a Bernoulli experiment

If one flips a coin with two sides, one will either obtain a head or a tail. For each flip, which side will be obtained is random. Such random experiments with two different mutually exclusive outcomes are called *Bernoulli experiments* or *Bernoulli trials.* If a coin is fair, we expect that the probability of observing a head is 0.5; that is if we repeatedly flip the coin many times, then both the head-side and the tail-side will be observed for about half the times. If the coin is not fair and is more favorable towards observing a head, say, then the probability of observing a head, denoted by p, is larger than 0.5. Many phenomena can be described using Bernoulli experiments.

Let the two possible outcomes of a Bernoulli experiment be labeled as 1 (success) and 0 (failure) with occurrence probabilities p and $1-p$, respectively. Let X be the variable that represents the outcome of a Bernoulli experiment, then the distribution of the random variable X is given in Table 6.5. The distribution is called a *Bernoulli distribution* and is usually denoted by $X \sim B(1, p)$, where $0 \leqslant p \leqslant 1$ is the probability of success.

Table 6.5 Bernoulli distribution

x	1 (success)	0 (failure)
probability $f(x)$	p	$1-p$

To estimate the probability of success p, a random sample $\{X_1, X_2, \ldots, X_n\}$ is collected with each of the X_i identically and independently distributed as $B(1, p)$. Let $\{x_1, x_2, \ldots, x_n\}$ be the observed outcomes of the random sample, where x_i equals to either 1 (success) or 0 (failure), then the likelihood function is given by

$$\begin{aligned} L^*(p) &= \prod_{i=1}^{n} p^{x_i}(1-p)^{(1-x_i)} \\ &= p^{\sum_{i=1}^{n} x_i}(1-p)^{(n-\sum_{i=1}^{n} x_i)}, \end{aligned}$$

and the log-likelihood is given by

$$L(p) = \ln L^*(p) = \left(\sum_{i=1}^{n} x_i \ln(p) + \left(n - \sum_{i=1}^{n} x_i \right) \ln(1-p) \right). \tag{6.12}$$

It can be shown (Exercise 6.6.1) that the maximum of $L(p)$ can be achieved when p is given by

$$\hat{p} = \frac{1}{n} \sum_{i=1}^{n} x_i = \bar{x}. \tag{6.13}$$

This is the maximum likelihood estimate of p, which is simply the proportion of success in the n trials. If we replace the actual estimate with the random variables that represent the sample, the maximum likelihood estimator of the success probability p in a Bernoulli experiment is given by

$$\hat{p} = \frac{1}{n} \sum_{i=1}^{n} X_i = \overline{X}. \tag{6.14}$$

It is worth noting that, we use the notation $\hat{p}$ to denote both the random variables (see (6.14)) and the estimated value (see (6.13)) of the success probability. This is a common practice in statistics literature to avoid the excess use of notation in presentation when the context is clear.

Exercise 6.6.1 Prove the result in (6.13).

Chapter 7

Continuous Random Variables and Related Concepts

7.1 Continuous random variables

Generalizing the concept of a discrete random variable to a random variable that can take on any value over a continuous range $\mathcal{R}$, we then have a *continuous random variable*. Associated with each continuous variable is a *probability density function.*

In Figure 6.1 of Section 6.2.2, we introduced the relative frequency histogram, which is a graphical representation of the probability function of a discrete random variable. Figure 7.1 shows that the analogous probability density function for a continuous random variable can be regarded as a smoothed continuous curve of the relative frequency histogram. We again use $f(x)$ to de-

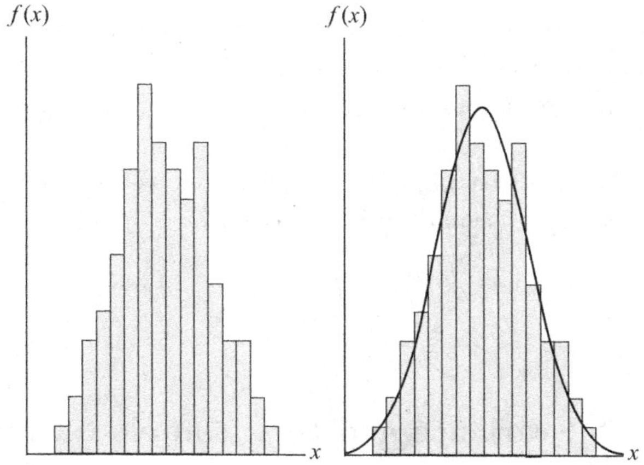

Fig. 7.1 Probability density function

note the probability density function of a continuous random variable. Similar to (6.1) and (6.2), a probability density function $f(x)$ possesses the following properties:

- $f(x) > 0 \quad \text{for all } x \in \mathcal{R}, \text{and}$ (7.1)
- $\int_{x \in \mathcal{R}} f(x)dx = 1.$ (7.2)

Unlike the probability function for a discrete random variable, the property $f(x) < 1$ is not a necessary condition for $f(x)$ to be a probability density function. For a continuous random variable, $f(x_0)$ at the point x_0 is not the probability of $X = x_0$. In effect, the probability that X equals to a specific value, say x_0, is always 0 because a tiny difference between x_0 and the value of X always exists. When X is a continuous random variable, the area under the curve defined by $f(x)$ and bounded by an interval provides the probability that X lies in the interval. That is, for given constants a and b, $\Pr(a < X < b) = \Pr(a < X \leqslant b) = \Pr(a \leqslant X < b) = \Pr(a \leqslant X \leqslant b) = \int_a^b f(x)dx$.

Corresponding to each probability density function $f(x)$ is the *distribution function* $F(x)$, which is given by

$$\Pr(X \leqslant x) = F(x) = \int_{-\infty}^{x} f(t)dt.$$

According to (7.2), $0 \leqslant F(x) \leqslant 1$.

7.2 Mathematical expectation for continuous random variables

We have explained the concept of mathematical expectation in the context of discrete random variables. From Figure 7.1, it is clear that the concept can be generalized to continuous random variables using integrations to replace summations.

Specifically, let $g(X)$ be a function of a continuous random variable X, then the *mathematical expectation* of $g(X)$, denoted by $\mathrm{E}(g(X))$, is given by

$$\mathrm{E}(g(X)) = \int_{x \in \mathcal{R}} g(x)f(x)dx. \tag{7.3}$$

7.3 Mean and variance and their sample estimates

Mathematical expectations are population-level quantities. Generalizing the discussion in Section 6.3, two typical population parameters are the *population mean* μ and the *population variance* σ^2. They respectively describe the location and dispersion of the population.

The *population mean* for a continuous variable is defined by

$$\mu = \mathrm{E}(X) = \int_{x \in \mathcal{R}} x f(x) dx, \tag{7.4}$$

and the *population variance* is defined by

$$\sigma^2 = \mathrm{Var}(X) = \mathrm{E}(X - \mu)^2 = \int_{x \in \mathcal{R}} (x - \mu)^2 f(x) dx. \tag{7.5}$$

Given a random sample $\{X_1, \ldots, X_n\}$, μ and σ^2 can be estimated, respectively, using the sample mean

$$\overline{X} = \frac{\sum_{i=1}^{n} X_i}{n}, \tag{7.6}$$

and the sample variance

$$S^2 = \frac{\sum_{i=1}^{n} (X_i - \overline{X})^2}{n - 1}. \tag{7.7}$$

From Exercise 7.3.1, the numerator in (7.7) can also be expressed as

$$\sum_{i=1}^{n} (X_i - \overline{X})^2 = \sum_{i=1}^{n} X_i^2 - n\overline{X}^2. \tag{7.8}$$

We will explain in Section 7.8 that $\overline{X}$ and S^2 are *unbiased* estimators. To better describe the concept of unbiased, it may be necessary to introduce some basic of the properties of mathematical expectations.

Exercise 7.3.1 Prove the results in (7.8).

7.4 Basic properties of expectations

We summarize some of the basic properties of mathematical expectation in this section. These properties are valid for both discrete and continuous random variables. They can be established by using the definition of mathematical expectation in (6.8) or (7.3) together with the property $\sum_{x \in \mathcal{R}} f(x) = 1$ (see (6.2)) or $\int_{x \in \mathcal{R}} f(x) = 1$ (see (7.2)) of the probability (density) function.

Let X be a random variable, c and $c_i, i = 1, \ldots, k$ be constants that are not random, and g and $g_i, i = 1, \ldots, k$ be functions of X. It can be shown that

$$\begin{aligned}
&\text{(i)}\ \ \mathrm{E}(c) = c \\
&\text{(ii)}\ \ \mathrm{E}(cg(X)) = c\mathrm{E}(g(X)) \\
&\text{(iia)}\ \ \mathrm{E}(cX) = c\mathrm{E}(X) \\
&\text{(iii)}\ \ \mathrm{E}(c + g(X)) = c + \mathrm{E}(g(X)) \\
&\text{(iiia)}\ \ \mathrm{E}(c + X) = c + \mathrm{E}(X) \\
&\text{(iv)}\ \ \mathrm{E}\left(\sum_{i=1}^{k} c_i g_i(X)\right) = \sum_{i=1}^{k} c_i \mathrm{E}(g_i(X)) \\
&\text{(iva)}\ \ \mathrm{E}\left(\sum_{i=1}^{k} c_i X\right) = \sum_{i=1}^{k} c_i \mathrm{E}(X)
\end{aligned} \tag{7.9}$$

Moreover, if we have m random variables $X_1, \ldots, X_m$, property (iv) can be generalized to

$$\mathrm{E}\left(\sum_{i=1}^{k} c_i g_i(X_1, \ldots, X_m)\right) = \sum_{i=1}^{k} c_i \mathrm{E}(g_i(X_1, \ldots, X_m)). \tag{7.10}$$

A special case of (7.10) is that

$$\mathrm{E}\left(\sum_{i=1}^{k} c_i X_i\right) = \sum_{i=1}^{k} c_i \mathrm{E}(X_i). \tag{7.11}$$

The results in (7.10) and (7.11) follow from those to be introduced in (8.11) and (8.12).

Exercise 7.4.1 Prove the results in (7.9) for both discrete and continuous random variables.

Exercise 7.4.2 Show that $\sigma^2 = \mathrm{E}(X - \mu)^2 = \mathrm{E}(X^2) - \mu^2$ for both discrete and continuous random variables.

7.5 Normal distribution

As previously explained in Section 6.5, the distribution of a population of interests is usually not known in practical analysis. To facilitate the drawing of statistical inferences from a sample to a population it is convenient to operate on the assumption that the distribution of the population belongs to a specific family, and that the normal distribution is a frequently used family.

The probability density function of a normal distribution is

$$f(x) = \frac{1}{\sigma\sqrt{2\pi}} \exp\left[-\frac{(x-\mu)^2}{2\sigma^2}\right], \quad -\infty < x < \infty. \tag{7.12}$$

The distribution is characterized by the two parameters μ and σ, where $-\infty < \mu < \infty$ and $0 < \sigma < \infty$. It can be shown using (7.12), (7.4) and (7.5) that $\mu = \mathrm{E}(X)$ is the expected value or mean of X, and $\sigma^2 = \mathrm{E}(X-\mu)^2 = \mathrm{Var}(X)$ is the variance of X. A random variable X that follows a normal distribution is usually denoted by $X \sim N[\mu, \sigma^2]$. Figure 7.2 depicts the probability density functions of several normal distributions with different combinations of mean μ and variance σ^2.

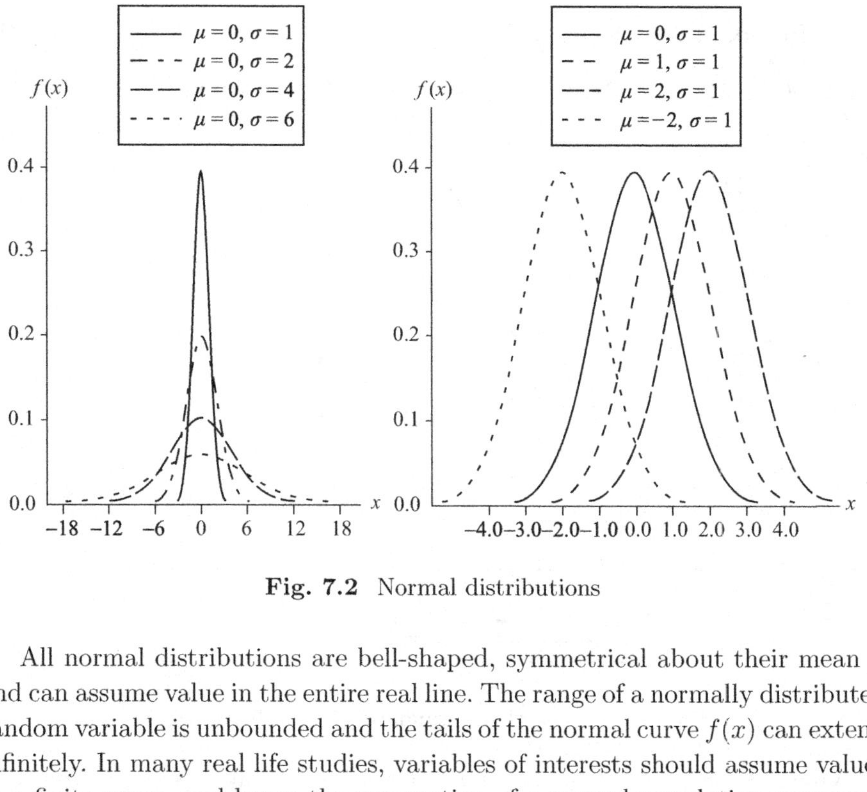

Fig. 7.2 Normal distributions

All normal distributions are bell-shaped, symmetrical about their mean μ and can assume value in the entire real line. The range of a normally distributed random variable is unbounded and the tails of the normal curve $f(x)$ can extend infinitely. In many real life studies, variables of interests should assume values in a finite range, and hence the assumption of a normal population may seem inappropriate. However, as the areas in the tails of a normal distribution are extremely small, such that the probability of the occurrence of tail values is negligible, the normal distribution is frequently employed in practical analysis to take advantage of its many nice and well established statistical properties.

If $\mu = 0$ and $\sigma^2 = 1$, then $X \sim N[0, 1]$ is called a *standard normal distribution.* The probability density function of a standard normal distribution is

$$f(x) = \frac{1}{\sqrt{2\pi}} \exp\left[-\frac{x^2}{2}\right], \quad -\infty < x < \infty. \tag{7.13}$$

From (7.13), one can easily find the area under the probability function that is bounded by an interval $[a, b]$. For example, if $X \sim N[0, 1]$, it can be easily shown that

$$\begin{aligned}
\Pr(-1 < X < 1) &= \int_{-1}^{1} f(x)dx = \int_{-1}^{1} \frac{1}{\sqrt{2\pi}} \exp\left(-\frac{x^2}{2}\right) dx = 0.683, \\
\Pr(-2 < X < 2) &= \int_{-2}^{2} f(x)dx = \int_{-2}^{2} \frac{1}{\sqrt{2\pi}} \exp\left(-\frac{x^2}{2}\right) dx = 0.954, \\
\Pr(-3 < X < 3) &= \int_{-3}^{3} f(x)dx = \int_{-3}^{3} \frac{1}{\sqrt{2\pi}} \exp\left(-\frac{x^2}{2}\right) dx = 0.997, \\
\Pr(-\infty < X < \infty) &= 1.
\end{aligned} \tag{7.14}$$

Figure 7.3 provides a graphical explanation of the results in (7.14).

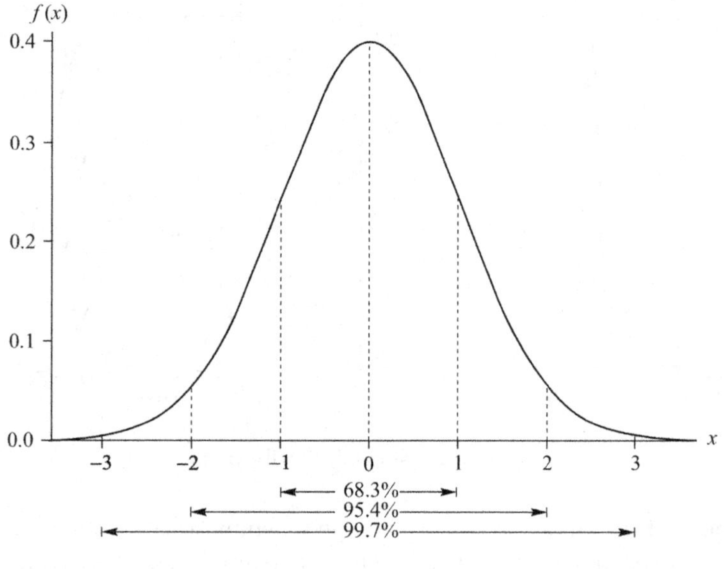

Fig. 7.3 Areas of the standard normal distribution

One nice property of the normal distribution is that if $X \sim N[\mu, \sigma^2]$ with density (7.12), then the density of the transformed random variable $Z = (X - \mu)/\sigma$ is given by (7.13) and hence, Z follows a standard normal distribution. That is

$$Z = \frac{X - \mu}{\sigma} \sim N[0, 1] \quad \text{if} \quad X \sim N[\mu, \sigma^2]. \tag{7.15}$$

Probabilities for standard normal distributions, such as those in (7.14), are readily available. The results in (7.15) enable these probabilities for standard normal distributions be used for other normal distributions with different μ and σ^2. For example, using (7.15) and the results in (7.14), we know that if $X \sim N[\mu, \sigma^2]$, then

$$\begin{aligned} \Pr(\mu - 1\sigma < X < \mu + 1\sigma) &= 0.683, \\ \Pr(\mu - 2\sigma < X < \mu + 2\sigma) &= 0.954, \\ \Pr(\mu - 3\sigma < X < \mu + 3\sigma) &= 0.997. \end{aligned} \tag{7.16}$$

Figure 7.4 provides a graphical explanation of the results in (7.16).

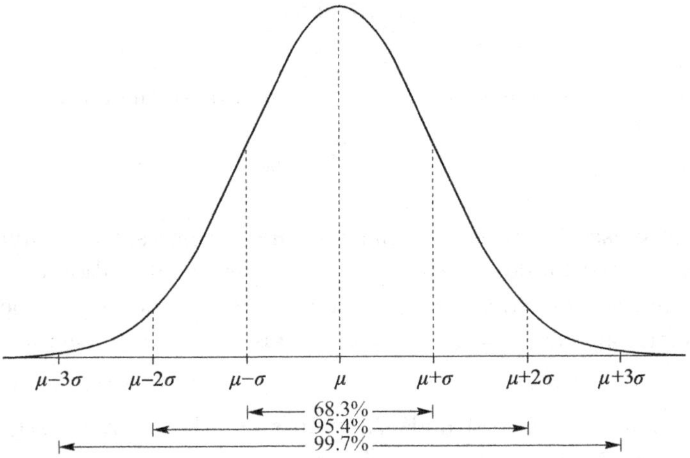

Fig. 7.4 Areas of the normal distributions

Statistical tables for standard normal distribution that provide information similar to those in (7.14) are widely available. These tables are in different formats, but all provide information in relation to *percentiles*. Specifically, for $p \in [0, 1]$, the 100p-th *percentile* is the value t such that

$$\Pr(X \leqslant t) = \int_{-\infty}^{t} f(x)dx = \int_{-\infty}^{t} \frac{1}{\sqrt{2\pi}} \exp\left(-\frac{x^2}{2}\right) dx = p. \tag{7.17}$$

Please refer to Figure 7.5 for a graphical representation of the meaning of a percentile.

For example, based on the symmetric properties of standard normal distribution, the 50-th percentile of the standard normal distribution is 0.

Percentile is a concept that can be applied to other distributions for discrete or continuous random variables. Statistical tables that provide various percentiles for other popular distributions such as the *t-distribution*, the *chi-squared distribution* and the *F distribution*, each with a different $f(x)$, are also widely available.

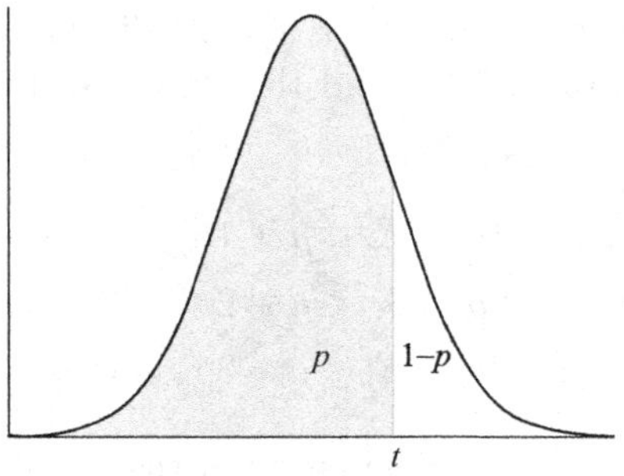

Fig. 7.5 Percentile

It is worth mentioning that the transformation of the form

$$Z = \frac{X - \mu}{\sigma} \tag{7.18}$$

is frequently used in statistical analysis, and not necessarily confined to normally distributed random variables. It measures the deviation of the X value from the mean using standard deviation as the unit of measurement. Similar transformations, say $z_i = (x_i - \bar{x})/s$, are also frequently used at the sample level.

Exercise 7.5.1 Use (7.3) and (7.12) to show that if $X \sim N(\mu, \sigma^2)$, then $\mathrm{E}(X) = \mu$ and $\mathrm{Var}(X) = \mathrm{E}(X - \mu)^2 = \sigma^2$.

Exercise 7.5.2 Use (7.12) and (7.13) to prove the result in (7.15).

7.6 Maximum likelihood estimation for continuous variables

The maximum likelihood method is introduced in Section 6.5 in the context of discrete random variables. It can be directly generalized to parameter estimation for distributions of continuous random variables. Let $\{X_1, \ldots, X_n\}$ be a random sample from a distribution with probability density function $f(x, \boldsymbol{\theta})$, where $\boldsymbol{\theta} = (\theta_1, \theta_2, \ldots, \theta_q)^T$ is a vector of the unknown parameters that characterize the distribution and $x_i, i = 1, \ldots, n$ is the observed values of $X_1, \ldots, X_n$, then the *maximum likelihood estimate* $\widehat{\boldsymbol{\theta}} = (\widehat{\theta}_1, \ldots, \widehat{\theta}_q)^T$ of $\boldsymbol{\theta}$ is obtained by maximizing the *likelihood function*

$$L^*(\boldsymbol{\theta}) = \prod_{i=1}^{n} f(x_i, \boldsymbol{\theta}), \tag{7.19}$$

or the log-likelihood

$$L(\boldsymbol{\theta}) = \ln L^*(\boldsymbol{\theta}) = \sum_{i=1}^{n} f(x_i, \boldsymbol{\theta}) \tag{7.20}$$

with respect to $\boldsymbol{\theta}$. The following section provides an example when the distribution is normal.

7.7 Maximum likelihood estimation for the parameters of normal distribution

Let $x_1, \ldots, x_n$ be the observed values of the random sample $X_1, \ldots X_n$ taken from a normal distribution $N[\mu, \sigma^2]$. From (7.12), the likelihood function is given by

$$L^*(\mu, \sigma^2) = \prod_{i=1}^{n} \frac{1}{\sqrt{2\pi\sigma^2}} \exp\left[-\frac{(x_i - \mu)^2}{2\sigma^2}\right], \tag{7.21}$$

and the log-likelihood function is

$$L(\mu, \sigma^2) = \log L^*(\mu, \sigma^2) = -\frac{n}{2}\log(2\pi\sigma^2) - \frac{1}{2\sigma^2}\sum_{i=1}^{n}(x_i - \mu)^2. \tag{7.22}$$

Taking partial derivatives of (7.22) with respect to μ and σ^2, respectively, and setting the partial derivatives to zero, we obtain the likelihood equations

$$\frac{\partial L(\mu, \sigma^2)}{\partial \mu} = \frac{1}{\sigma^2}\sum_{i=1}^{n}(x_i - \mu) = 0, \tag{7.23}$$

and

$$\frac{\partial L(\mu, \sigma^2)}{\partial \sigma^2} = \frac{-n}{2\sigma^2} + \frac{1}{2(\sigma^2)^2}\sum_{i=1}^{n}(x_i - \mu)^2 = 0. \tag{7.24}$$

Solving equations (7.23) and (7.24), we obtain the maximum likelihood estimates

$$\widehat{\mu}_{ML} = \frac{\sum_{i=1}^{n} x_i}{n} = \bar{x}, \tag{7.25}$$

and

$$\widehat{\sigma}^2_{ML} = \frac{\sum_{i=1}^{n}(x_i - \bar{x})^2}{n}. \tag{7.26}$$

If we replace the observed values $x_1, \ldots, x_n$ with the corresponding random variables $X_1, \ldots, X_n$, we obtain the following maximum likelihood estimators for the normal distribution parameters:

$$\widehat{\mu}_{ML} = \frac{\sum_{i=1}^{n} X_i}{n} = \overline{X}, \tag{7.27}$$

and

$$\widehat{\sigma}^2_{ML} = \frac{\sum_{i=1}^{n} (X_i - \overline{X})^2}{n}. \tag{7.28}$$

To avoid the excess use of notation, we use the same notation to denote both the estimator and the estimate. For example, $\widehat{\mu}_{ML}$ in (7.25) represents an estimate, whereas (7.27) it represents an estimator.

Moreover, the maximum likelihood estimator $\widehat{\sigma}^2_{ML}$ of σ^2 is not the same as the sample variance S^2 in (7.7), which is also used to estimate the population variance σ^2. The denominators that are respectively n and $n-1$ are different. S^2 is an unbiased estimator, but $\widehat{\sigma}^2_{ML}$ is not unbiased in small a sample. We explain this in the next section.

7.8 Sampling distribution

Let $\{X_1, X_2, \ldots, X_n\}$ be a random sample from a population with density $f(x)$. Every sample statistic that is a function of the X_i's is a random variable, and has a distribution that depends on the original distribution of the X_i's. The distribution of a sample statistic is called a *sampling distribution*. We use Table 7.1 to illustrate the intuitive meaning of a sampling distribution, particularly the sampling distributions of the sample mean $\overline{X}$, the sample variance $S^2 = \sum_{i=1}^{n}(X_i - \overline{X})^2/(n-1)$ and $\widehat{\sigma}^2_{ML} = \sum_{i=1}^{n}(X_i - \overline{X})^2/n$.

In practical analysis, we estimate the population mean μ and population variance σ^2 by drawing one single random sample $\{X_1, X_2, \ldots, X_n\}$ from the population and computing the sample mean $\overline{X}$, the sample variance S^2 or $\widehat{\sigma}^2_{ML}$. Imagine if one were to repeat the process many times, one would obtain many different samples, such as those illustrated in Table 7.1, and each of which would give a different set of estimates. For example, the m-th sample $\{X_1^{(m)}, \ldots, X_n^{(m)}\}$ gives the sample mean $\overline{X}_{(m)}$, sample variance $S^2_{(m)}$ and ML estimate $\widehat{\sigma}^2_{ML(m)}$. Consider the sample mean and let the many different means obtained from different samples be denoted by $\overline{X}_{(1)}, \overline{X}_{(2)}, \ldots$ then the collection $\{\overline{X}_{(1)}, \overline{X}_{(2)}, \ldots\}$ can be regarded as a population of $\overline{X}$. Associated with $\overline{X}$

is a distribution that provides information on the probability of the occurrence of different $\bar{x}$ values. Such a distribution for a sample statistic is called a *sampling distribution.* The sampling distribution depends on the distribution of the population from which the sample is drawn, in addition to the sample size. One can see that if the dispersion of the original population is small or when the sample size is large, then the values of the many $\overline{X}_{(1)}, \overline{X}_{(2)}, \ldots$ may be quite close, which renders the dispersion of the distribution for $\overline{X}$ very small.

Similar to $\overline{X}$, the collections $\{S^2_{(1)}, S^2_{(2)}, \ldots\}$ and $\{\widehat{\sigma}^2_{ML(1)}, \widehat{\sigma}^2_{ML(2)}, \ldots\}$ are regarded as the sampling distributions of S^2 and $\widehat{\sigma}^2_{ML}$, respectively.

Table 7.1 The meaning of sampling distributions and unbiased estimators

Samples taken from the population represented by the random variable X, with $E(X) = \mu$, $Var(X) = \sigma^2$

Sample	Size n sample	$\overline{X}$	S^2	$\widehat{\sigma}^2_{ML}$
1	$X_1^{(1)}, \ldots, X_n^{(1)}$	$\overline{X}_{(1)}$	$S^2_{(1)}$	$\widehat{\sigma}^2_{ML(1)}$
2	$X_1^{(2)}, \ldots, X_n^{(2)}$	$\overline{X}_{(2)}$	$S^2_{(2)}$	$\widehat{\sigma}^2_{ML(2)}$
$\vdots$	$\vdots$	$\vdots$	$\vdots$	$\vdots$
m	$X_1^{(m)}, \ldots, X_n^{(m)}$	$\overline{X}_{(m)}$	$S^2_{(m)}$	$\widehat{\sigma}^2_{ML(m)}$
$\vdots$	$\vdots$	$\vdots$	$\vdots$	$\vdots$
Sampling distribution		$\{\overline{X}_{(1)}, \overline{X}_{(2)}, \ldots\}$	$\{S^2_{(1)}, S^2_{(2)}, \ldots\}$	$\{\widehat{\sigma}^2_{ML(1)}, \widehat{\sigma}^2_{ML(2)}, \ldots\}$
Mean of sampling distribution		$\mathrm{E}(\overline{X}) = \mu$, i.e., the average of $\{\overline{X}_{(1)}, \overline{X}_{(2)}, \ldots\}$ is μ	$\mathrm{E}(S^2) = \sigma^2$, i.e., the average of $\{S^2_{(1)}, S^2_{(2)}, \ldots\}$ is σ^2	$\mathrm{E}(\widehat{\sigma}^2_{ML}) = \sigma^2$, i.e., the average of $\{\widehat{\sigma}^2_{ML(1)}, \widehat{\sigma}^2_{ML(2)}, \ldots\}$ is σ^2 if n is large

We show in (6.4) and (7.4), that for each random variable we can find its expected value, which can be interpreted as the mean of the distribution. Because a sample statistic is a random variable, we can find its expected value or, equivalently, the mean of the sampling distribution. Such an expected value can usually be obtained by making use of the properties of the population from which the sample is drawn. Let $\overline{X}$ be the sample mean of the random sample $\{X_1, \ldots, X_n\}$, which is taken from a population with mean μ, then $\mathrm{E}(X_i) = \mu$ for all i. To find the mean of the sampling distribution of $\overline{X}$, (7.11) shows that

$$\mathrm{E}(\overline{X}) = \mathrm{E}\left(\frac{1}{n}\sum_{i=1}^{n} X_i\right) = \sum_{i=1}^{n}\frac{1}{n}\mathrm{E}(X_i) = \frac{1}{n}\sum_{i=1}^{n}\mu = \frac{n\mu}{n} = \mu. \qquad (7.29)$$

Equation (7.29) says that the mean of the population of $\overline{X}$ equals to μ, which is the true parameter that we intend to estimate for the original population of interest. As illustrated in Table 7.1, this means that if we draw many samples and then compute the sample mean for each sample, the average of the many sample means $\{\overline{X}_{(1)}, \overline{X}_{(2)}, \ldots\}$ equals to μ. The property in (7.29) is a desirable property in estimation, and is called that $\overline{X}$ is an *unbiased estimator* of μ.

In general, we call an estimator $\widehat{\theta}$ an *unbiased estimator* of a population parameter θ if

$$\mathrm{E}(\widehat{\theta}) = \theta. \tag{7.30}$$

The sample mean $\overline{X}$ is commonly used to estimate the population mean μ. In practical analysis, we collect only one instead of many samples and hence, we have only one rather than many sample means. We have no idea whether the one single sample mean is close to μ or not, but we are happy to use it to estimate the population mean μ because we understand that our estimator is unbiased. We know that if we can repeat the process many times then, on average, we can produce an accurate estimate.

In Section 6.3.4, we stated that the sample variance

$$S^2 = \sum_{i=1}^{n} (X_i - \overline{X})^2/(n-1) \tag{7.31}$$

with $n-1$ rather than n in its denominator is a preferred estimator of $\sigma^2 = \mathrm{E}(X-\mu)^2$. This is because S^2, with $n-1$ in its denominator, is an unbiased estimator for σ^2. That is, $\mathrm{E}(S^2) = \sigma^2$. As illustrated in Table 7.1, this means that the average of the many $S^2_{(1)}, S^2_{(2)}, \ldots$ that are obtained from different samples, each computed using (7.31), equals to σ^2. However, $\widehat{\sigma}^2_{ML} = \sum_{i=1}^{n}(X_i - \overline{X})^2/n$ is not an unbiased estimator and $E(\widehat{\sigma}^2_{ML})$ is not equal to σ^2 when the sample size n is small. When the sample size n is large, the difference between n and $n-1$ is small and hence, $\widehat{\sigma}^2_{ML}$ is close to S^2 and both are unbiased estimators of σ^2.

Different sample statistics are used to estimate different population parameters, and unbiased is a desirable property for a sample statistic. The maximum likelihood approach is a popular method for finding parameter estimates and that all maximum likelihood estimators are *asymptotically unbiased.* Specifically, based on a sample of size n collected from a distribution $f(x) = f(x, \theta)$ with unknown parameter θ, if the sample statistic Z_n as an estimate of the unknown parameter θ is obtained using the maximum likelihood method, then it can be shown that when sample size is large, $\mathrm{E}(Z_n) = \theta$ when n is sufficiently large. In particular, $\widehat{\sigma}^2_{ML}$ is unbiased when n is large, although it is not unbiased when sample size n is small.

Chapter 8

Bivariate and Multivariate Distribution

The discussion in Chapters 6 and 7 addresses univariate distribution with a random variable that measures one single characteristic. To analyze two or more characteristics, we need to consider bivariate or multivariate distributions with two or more random variables.

8.1 Bivariate distribution for discrete random variables

8.1.1 Joint probability function

Let (X, Y) be a pair of discrete random variables that can assume values $(x, y) \in \mathcal{R}$, where $\mathcal{R}$ is a set with a finite or countable number of elements. The function $f(x, y) = \Pr(X = x, Y = y)$ that specifies the probability of the occurrence of (x, y), is called the *joint probability function.*

Similar to the univariate case (see (6.1) and (6.2)), the joint probability function for discrete random variables possesses the following properties:

- $0 < f(x, y) \leqslant 1$ for all $(x, y) \in \mathcal{R}$, and
- $\sum_{(x,y)\in\mathcal{R}} f(x, y) = 1.$ (8.1)

Example 8.1.1 (Bivariate distribution) Consider the household size problem in a small town with 100,000 households in Example 6.2.1. Suppose that in addition to household size, we also have information on the number of male heads of households, which is given in Table 8.1. If we randomly select a household and let X be the household size and Y its number of male heads, then the joint probability of observing a pair of specific (x, y) values can be derived and is given in Table 8.2. For example, the probability of observing a household of size 4 with 2 male heads is 0.09.

Table 8.1 Household size and number of male heads

	No. of male heads Y					
Household size X	0	1	2	3	4	Total
1	5,000	5,000	0	0	0	10,000
2	4,000	12,000	4,000	0	0	20,000
3	5,000	15,000	15,000	5,000	0	40,000
4	3,000	7,500	9,000	7,500	3,000	30,000
Total	17,000	39,500	28,000	12,500	3,000	100,000

Table 8.2 Joint and marginal probabilities

	No. of male heads Y					
Household size X	0	1	2	3	4	Total
1	0.05	0.05	0	0	0	0.1
2	0.04	0.12	0.04	0	0	0.2
3	0.05	0.15	0.15	0.05	0	0.4
4	0.03	0.075	0.09	0.075	0.03	0.3
Total	0.17	0.395	0.28	0.125	0.03	1.0

8.1.2 Marginal probability function

In Table 8.2, the row and column totals are, respectively, the probabilities of observing various household sizes and male heads. For example, the probability of observing 3 male heads is 0.125, which is the total of the entries in the column specifying that $Y = 3$. The probability of observing a household size of 2 is 0.2, which is the total of the entries in the row specifying that $X = 2$. These probabilities are called marginal probabilities.

Specifically, let $f(x, y)$ be the joint probability function for X and Y, then the *marginal probability functions* of X and Y are respectively given by

$$f_X(x) = f(x) = \sum_y f(x, y) \quad \text{and} \quad f_Y(y) = f(y) = \sum_x f(x, y), \tag{8.2}$$

where the summations are taken over all possible values of Y and X, respectively. The subscript X in $f_X(x)$ or Y in $f_Y(y)$ is usually omitted when the context is clear, and the marginal probability function is denoted by $f(x)$ or $f(y)$. In effect the marginal probability function for a variable is also its probability function when the variable is being analyzed on its own as a univariate random variable.

8.1.3 Conditional probability function

Table 8.2 shows that the probability of observing a household with 1 male head is 0.395. If we have the information that the size of the household is 2, then Table 8.1 reveals that the probability of observing 1 male head becomes

$$\begin{aligned}&\Pr(1 \text{ male head given household size } {=}2)\\ &= \Pr(Y=1|X=2) = \frac{12,000}{20,000} = 0.6. \end{aligned} \tag{8.3}$$

Such probabilities are given in Table 8.3 and are called the *conditional probabilities.*

Table 8.3 Conditional probabilities $f(y|x) = \Pr(Y = y|X = x)$

	No. of male heads Y					
Household size X	0	1	2	3	4	Total
1	0.5	0.5	0	0	0	1.0
2	0.2	0.6	0.2	0	0	1.0
3	0.125	0.375	0.375	0.125	0	1.0
4	0.1	0.25	0.3	0.25	0.1	1.0

Conditional probabilities, such as those in (8.3), can be obtained if the joint probabilities of the two variables are available. For example, from Table 8.2, the conditional probability of 1 male head, given a household of size 2 is

$$\Pr(Y=1|X=2) = \frac{12,000}{20,000} = \frac{\dfrac{12,000}{100,000}}{\dfrac{20,000}{100,000}} = \frac{0.12}{0.2} = \frac{\Pr(X=2 \text{ and } Y=1)}{\Pr(X=2)}.$$

This is the ratio of the joint probability of household size 2 and 1 male head (=0.12) to the marginal probability of household size 2 (=0.2).

In general, let $f(x, y)$ be the joint probability function of X and Y, and $f(x) = f_X(x)$ and $f(y) = f_Y(y)$ be the marginal probability function of X and Y respectively, then the *conditional probability function* of Y, given $X = x$, is defined by

$$f(y|x) = \Pr(Y=y|X=x) = \frac{f(x,y)}{f_X(x)} = \frac{f(x,y)}{f(x)}, \tag{8.4}$$

where $f_X(x) = f(x) > 0$. Similarly, the conditional probability function of X given $Y = y$ is defined by

$$f(x|y) = \Pr(X=x|Y=y) = \frac{f(x,y)}{f_Y(y)} = \frac{f(x,y)}{f(y)}, \tag{8.5}$$

where $f_Y(y) = f(y) > 0$.

Exercise 8.1.2 Use (8.5) and the joint and marginal probabilities in Table 8.2 to find the conditional probabilities $f(x|y) = \Pr(X = x|Y = y)$.

8.2 Bivariate distribution for continuous random variables

If the random variables X and Y are continuous, such as when $\mathcal{R}$ is a subset of $\mathbb{R}^2$, then the *joint probability density function* $f(x, y)$ for (X, Y) has the following properties:

- $f(x, y) > 0$
- $\int_x \int_y f(x, y) dy dx = 1$, and (8.6)
- $\Pr(a_1 \leqslant X \leqslant a_2 \text{ and } b_1 \leqslant Y \leqslant b_2) = \int_{a_1}^{a_2} \int_{b_1}^{b_2} f(x, y) dy dx.$

Generalizing (8.2), the *marginal probability density functions* (or, equivalently, the probability density functions) for X and Y are respectively given by

$$f_X(x) = f(x) = \int_y f(x, y) dy \quad \text{and} \quad f_Y(y) = f(y) = \int_x f(x, y) dx, \tag{8.7}$$

where the integrations are taken over the ranges of Y and X, respectively.

Moreover, the same is the case in (8.4) for discrete random variables, such that the conditional probability density functions of X given $Y = y$, and those of Y given $X = x$ are, respectively,

$$\begin{aligned} f(x|y) &= \frac{f(x, y)}{f_Y(y)} = \frac{f(x, y)}{f(y)}, \quad f(y) > 0; \quad \text{and} \\ f(y|x) &= \frac{f(x, y)}{f_X(x)} = \frac{f(x, y)}{f(x)}, \quad f(x) > 0. \end{aligned} \tag{8.8}$$

8.3 Mathematical expectations

8.3.1 Mathematical expectations for the functions of two random variables

Similar to the univariate case, we can define the mathematical expectation for a function g of two random variables, X and Y. For discrete random variables, we have

$$\mathrm{E}(g(X,Y)) = \sum_x \sum_y g(x,y)f(x,y), \tag{8.9}$$

and for continuous random variables, we have

$$\mathrm{E}(g(X,Y)) = \int_x \int_y g(x,y)f(x,y)dydx, \tag{8.10}$$

where the summations or the integrations are taken over the ranges of the values of X and Y, respectively.

Example 8.3.1 Let $g(X,Y) = X$, then we obtain the population mean μ_X of X. That is, if X is discrete,

$$\mu_X = \mathrm{E}(X) = \sum_x \sum_y xf(x,y) = \sum_x xf(x),$$

and if X is continuous

$$\mu_X = \mathrm{E}(X) = \int_x \int_y xf(x,y)dydx = \int_x xf(x)dx.$$

Example 8.3.2 Let $g(X,Y) = (X - \mu_X)^2$, then we obtain the population variance σ_X^2 of X. That is, if X is discrete,

$$\sigma_X^2 = \mathrm{E}(X - \mu_X)^2 = \sum_x \sum_y (x - \mu_X)^2 f(x,y) = \sum_x (x - \mu_X)^2 f(x),$$

and if X is continuous,

$$\sigma_X^2 = \mathrm{E}(X - \mu_X)^2 = \int_x \int_y (x - \mu_X)^2 f(x,y)dydx = \int_x (x - \mu_x)^2 f(x)dx.$$

It can be shown that (see Exercise 8.3.3) for given constants a and b and two functions $g_1(X,Y)$ and $g_2(X,Y)$ of X and Y, we have

$$\mathrm{E}(ag_1(X,Y) + bg_2(X,Y)) = a\mathrm{E}(g_1(X,Y)) + b\mathrm{E}(g_2(X,Y)). \tag{8.11}$$

A special case of (8.11) is

$$\mathrm{E}(aX + bY) = a\mathrm{E}(X) + b\mathrm{E}(Y). \tag{8.12}$$

Exercise 8.3.3 Use (8.9) or (8.10) to prove the results in (8.11) and (8.12).

8.4 Covariance and correlation

8.4.1 Sample covariance and correlation

Studies of the two characteristics represented by the pair of random variables (X, Y) are usually based on a random sample of size n, which is represented by $\{(X_i, Y_i), i = 1, \ldots, n\}$. The sample means, $\overline{X}$ and $\overline{Y}$, and the sample variances, S_X^2 and S_Y^2, are sample statistics that describe the univariate properties of X and Y. The *correlation coefficient* is a popular measure for examine the association or the relationship between the two variables, particularly when both X and Y are continuous in scale.

The *sample correlation coefficient* for $\{(X_i, Y_i)\}, i = 1, \ldots, n\}$ is given by

$$r = \frac{\sum_{i=1}^{n}(X_i - \overline{X})(Y_i - \overline{Y})}{\sqrt{\sum_{i=1}^{n}(X_i - \overline{X})^2}\sqrt{\sum_{i=1}^{n}(Y_i - \overline{Y})^2}} = \frac{\sum_{i=1}^{n}(X_i - \overline{X})(Y_i - \overline{Y})}{(n-1)S_X S_Y}, \tag{8.13}$$

where $S_X = \sqrt{S_X^2}$ and $S_Y = \sqrt{S_Y^2}$. This is a measure of the strength of the linear relationship between two variables. The quantity

$$S_{XY} = \sum_{i=1}^{n} \frac{(X_i - \overline{X})(Y_i - \overline{Y})}{n-1} \tag{8.14}$$

is the *sample covariance* between X and Y.

From Exercise 8.4.1, the numerator of r and S_{XY} can be expressed as

$$\sum_{i=1}^{n}(X_i - \overline{X})(Y_i - \overline{Y}) = \sum_{i=1}^{n} X_i Y_i - n\overline{X}\,\overline{Y}. \tag{8.15}$$

If most of the data points in the data set with larger X values are accompanied by larger Y values, then the quantities $(X_i - \overline{X})(Y_i - \overline{Y})$ for most of the data points are positive and hence, S_{XY} as an aggregated quantity is also positive. However, if most of the data points in the data set with larger X values are accompanied by smaller Y values, then the quantities $(X_i - \overline{X})(Y_i - \overline{Y})$ for most of the data points are negative and hence, the aggregated quantity S_{XY} is negative. For data sets in which X and Y have no relationship, such that the larger or smaller X values pair up with larger or smaller Y values in a random manner, then the values $(X_i - \overline{X})(Y_i - \overline{Y})$ are positive for some data points and negative for others and hence, S_{XY} is close to zero. The magnitude of S_{XY} depends on the unit of measurement for X and Y and is difficult to interpret. To facilitate interpretation, (8.13) is used to produce the correlation coefficient r, which assumes value in the interval $[-1, 1]$. Figure 8.1 shows the *scatter plots* of several data sets with different correlation coefficients. A scatter plot is simply the plot of the data points in the xy-plane. As S_X and S_Y are positive, we know that:

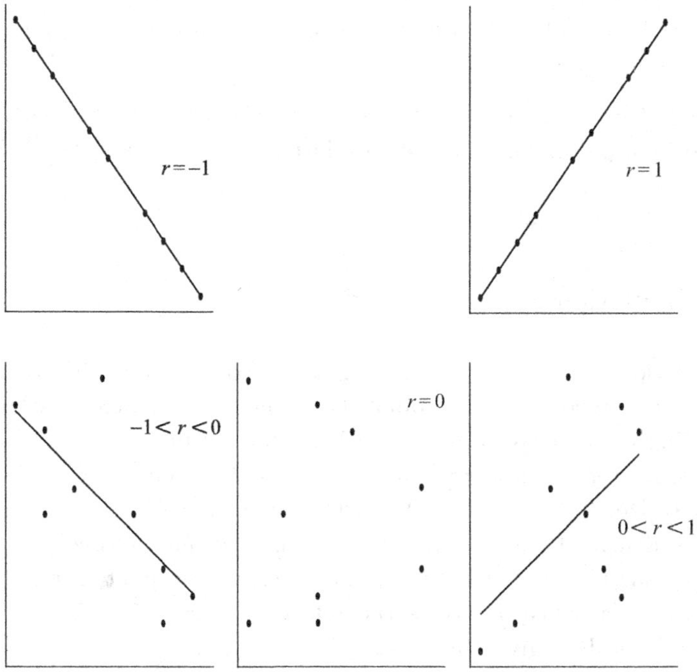

Fig. 8.1 Data sets with different correlations

- $r = 0$ indicates that the two variables have no linear relationship.
- $r > 0$ indicates a positive linear relationship. The two variables are likely to move along the same direction and larger X values tend to pair up with larger Y values.
- $r = 1$ represents a perfect positive relationship in which all data points lie on a straight line with a positive slope (see Exercise 8.4.2).
- $r < 0$ indicates a negative linear relationship. The two variables are likely to move along opposite directions and larger X values tend to pair up with smaller Y values.
- $r = -1$ represents a perfect negative relationship in which all data points lie on a straight line with a negative slope (see Exercise 8.4.2).

It is worth noting that r measures the strength of the linear relationship between two variables. It is possible that even when X and Y have a strong nonlinear relationship, the correlation coefficient is close to zero (see Exercise

8.4.3). Moreover, high correlation does not imply a casual relationship. That is, even if the correlation coefficient between X and Y is large, one cannot conclude that an increase in Y is due to an increase in X.

Exercise 8.4.1 Prove the result in (8.15).

Exercise 8.4.2 Use (8.13) to show that if all data points lie on a straight line with $Y_i = \alpha + \beta X_i$ for some α and β, then $r = -1$ if $\beta < 0$, $r = 1$ if $\beta > 0$ and $r = 0$ if $\beta = 0$.

Exercise 8.4.3 Illustrate that if a set of data points $\{(X_i, Y_i)\}$ can be closely resembled by a parabola, then the correlation coefficient r in (8.13) is close to zero.

8.4.2 Population covariance and correlation

The correlation coefficient r in (8.13) is a sample statistic, the corresponding population parameter is the population correlation coefficient, which can be defined using the concept of mathematical expectation.

The parameters μ_X and σ_X^2 characterize the distribution of X while μ_Y and σ_Y^2 characterize that of Y. Let (X, Y) be a pair of random variables with the probability density function $f(x, y)$, then the *population covariance* between X and Y, usually denoted by σ_{XY}, characterizes the joint distribution of X and Y. The covariance describes the relationship between X and Y at the population level. It is given by

$$\sigma_{XY} = \mathrm{Cov}(X, Y) = \mathrm{E}((X - \mu_X)(Y - \mu_Y)) = \sum_x \sum_y (x - \mu_X)(y - \mu_Y) f(x, y) \tag{8.16}$$

if X and Y are discrete, and

$$\sigma_{XY} = \mathrm{Cov}(X, Y) = \mathrm{E}((X - \mu_X)(Y - \mu_Y)) = \int_x \int_y (x - \mu_X)(y - \mu_Y) f(x, y) dy dx \tag{8.17}$$

if X and Y are continuous. σ_{XY} is a population analog of the sample covariance S_{XY} in (8.14) and has similar interpretations. If the pairs of values in the population of (X, Y) are such that larger than average X values tend to pair up with larger than average Y values, then σ_{XY} is positive. If the larger than average X values tend to pair up with smaller than average Y values, then σ_{XY} is negative. If there exists no special pattern, then σ_{XY} is close to zero.

Using (8.11) and the properties for mathematical expectations outlined in (7.9), it can be shown that

$$\sigma_{XY} = \mathrm{Cov}(X, Y) = \mathrm{E}((X - \mu_X)(Y - \mu_Y)) = \mathrm{E}(XY) - \mu_X \mu_Y. \tag{8.18}$$

Moreover, the covariance of a random variable with itself is the variable's variance; that is,

$$\sigma_{XX} = \text{Cov}(X, X) = \text{E}((X - \mu_X)^2) = \text{E}(X^2) - \mu_X^2 = \text{Var}(X) = \sigma_X^2. \quad (8.19)$$

Definition 8.4.4 The population correlation coefficient between the two random variables X and Y, usually denoted by ρ_{XY} or ρ, is defined as

$$\rho_{XY} = \rho = \frac{\text{E}[(X - \mu_X)(Y - \mu_Y)]}{\sigma_X \sigma_Y} = \frac{\text{Cov}(X, Y)}{\sigma_X \sigma_Y} = \frac{\sigma_{XY}}{\sigma_X \sigma_Y},$$

where $\sigma_X = \sqrt{\sigma_X^2}$ and $\sigma_Y = \sqrt{\sigma_Y^2}$. It can be shown that $-1 \leqslant \rho \leqslant 1$. Moreover, as both σ_X and σ_Y are positive, the interpretations of ρ_{XY} are similar to those for σ_{XY} and the sample-level correlation coefficient r. In effect, the population correlation coefficient ρ is a population analog of the sample correlation coefficient r in (8.13), which can be interpreted in the same way as that of r. For example, ρ_{XY} is positive if larger than average X values tend to pair up with larger than average Y values. The use of the correlation coefficient rather than covariance facilitates interpretation because it assumes value in the interval $[-1, 1]$.

Exercise 8.4.5 Show that if $Y = \alpha + \beta X$, then $\rho_{XY} = 1$ if $\beta > 0$ and $\rho_{XY} = -1$ if $\beta < 0$.

Exercise 8.4.6 Show that if $X = c$, where c is a constant and not a random variable, then $\text{Cov}(X, Y) = 0$.

8.4.3 Conditional expectations

For the household example in Table 8.1, the marginal distribution of the variable Y = *the number of male heads* is given by the column totals in Table 8.2, namely 0.17, 0.395, 0.28, 0.125 and 0.03 for $Y = 0, 1, 2, 3$ and 4, respectively. Based on the marginal distribution, the average number of male heads of all the households in the small town can be obtained by using the mathematical expectation $\mu_Y = \text{E}(Y)$, as follows:

$$0.17 \times 0 + 0.395 \times 1 + 0.28 \times 2 + 0.125 \times 3 + 0.03 \times 4 = 1.45. \quad (8.20)$$

Similarly, the number of male heads for a randomly drawn household among the households with a specific household size x is also a random variable. It is denoted by $Y|X = x$ and its conditional probability function $f(y|x)$ is available in Table 8.3. Similar to any other random variables, one can use these conditional probabilities to find the average number of male heads of all the households for a given household size x using the concept of mathematical expectation discussed in Section 6.3.1.

In general, the *conditional mean* of Y, given $X = x$, is defined by

$$\mu_{Y|X=x} = \mu_{Y|x} = \mathrm{E}(Y|X = x) = \mathrm{E}(Y|x) = \sum_y yf(y|x)$$

for discrete random variables, or

$$\mu_{Y|X=x} = \mu_{Y|x} = \mathrm{E}(Y|X = x) = \mathrm{E}(Y|x) = \int_y yf(y|x)dy$$

for continuous random variables.

Example 8.4.7 From the conditional probabilities $f(y|x)$ in Table 8.3, the average number of male heads of size 2 households is given by

$$0.2 \times 0 + 0.6 \times 1 + 0.2 \times 2 = 1.$$

This average number is different from the average number of male heads of all households in the small town 1.45 in (8.20), which is computed using marginal probabilities.

Conditional mathematical expectation for other functions $g(Y)$, given $X = x$, can be defined and interpreted in a similar manner. That is,

$$\mathrm{E}(g(Y)|X = x) = \sum_y g(y)f(y|x)$$

for discrete random variables, and

$$\mathrm{E}(g(Y)|X = x) = \int_y g(y)f(y|x)dy$$

for continuous random variables. For example, the *conditional variance* of Y, given $X = x$, is defined by

$$\sigma^2_{Y|X=x} = \mathrm{E}[(Y - \mu_{Y|x})^2|X = x] = \sum_y (y - \mu_{Y|x})^2 f(y|x)$$

for discrete random variables, and

$$\sigma^2_{Y|X=x} = \mathrm{E}[(Y - \mu_{Y|x})^2|X = x] = \int_y (y - \mu_{Y|x})^2 f(y|x)dy$$

for continuous random variables.

Conditional mathematical expectations for functions of $g(X)$ of X, given $Y = y$, can be defined and interpreted similarly.

The concept of conditional expectation is useful for formulating the simple linear regression model discussed in Chapter 9.

Exercise 8.4.8 In the small town example in Table 8.1, find the average household size for those households with 2 male heads.

8.5 Bivariate normal distribution

When we are interested in studying two characteristics represented by the pair of variables (X, Y), but the distribution of (X, Y) is not known, we may need to impose a distributional assumption on (X, Y) to facilitate the analysis. This is similar to what we usually do when we analyze univariate variable (see Section 6.5). In many applications, when X and Y are continuous-scale measurements, it is sensible to assume that the pair of variables (X, Y) follows a bivariate normal distribution. The probability density function for a bivariate normal distribution is given by

$$\begin{aligned} & f(x, y) \\ & = \frac{1}{2\pi\sigma_X\sigma_Y\sqrt{1-\rho_{XY}^2}}\exp\left\{\frac{-1}{2(1-\rho_{XY}^2)}\cdot\right. \\ & \left.\left[\left(\frac{x-\mu_X}{\sigma_X}\right)^2 - 2\rho_{XY}\left(\frac{x-\mu_X}{\sigma_X}\right)\left(\frac{y-\mu_Y}{\sigma_Y}\right) + \left(\frac{y-\mu_Y}{\sigma_Y}\right)^2\right]\right\}. \end{aligned} \tag{8.21}$$

The function $f(x, y)$ can be plotted with respect to (x, y) in a three-dimensional space. Generalizing Fig. 7.2 from a univariate to a bivariate case, the bivariate normal probability function $f(x, y)$ looks like a small hill located on the (x, y)-plane, such as the one in Fig. 8.2. The volume of the hill over a subset of the (x, y)-plane represents the probability of occurrence of the (x, y) values in the subset.

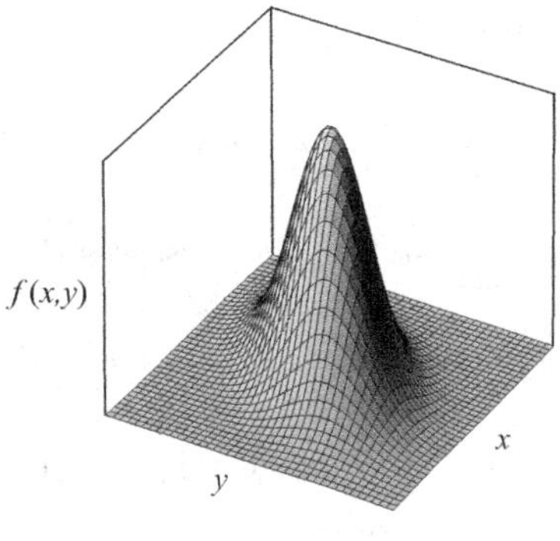

Fig. 8.2 Bivariate normal distribution

Bivariate normal distributions are characterized and determined by the parameters μ_X, μ_Y, σ_X^2, σ_Y^2 and $\rho_{XY} = \sigma_{XY}/\sigma_X\sigma_Y$. They are respectively the mean of X, the mean of Y, the variance of X, the variance of Y, and the correlation between X and Y. The bivariate normal distribution is usually denoted by $N[\boldsymbol{\mu}, \boldsymbol{\Sigma}]$; that is,

$$\begin{pmatrix} X \\ Y \end{pmatrix} \sim N\left[\boldsymbol{\mu} = \begin{pmatrix} \mu_X \\ \mu_Y \end{pmatrix}, \boldsymbol{\Sigma} = \begin{pmatrix} \sigma_X^2 & \sigma_{XY} \\ \sigma_{XY} & \sigma_Y^2 \end{pmatrix}\right], \tag{8.22}$$

where $\boldsymbol{\mu}$ is the mean vector and $\boldsymbol{\Sigma}$ is the *variance covariance matrix* or the *covariance matrix*. The elements in $\boldsymbol{\mu}$ are the population means or the expected values of X and Y, respectively. The diagonal elements of $\boldsymbol{\Sigma}$ are the variances of X and Y, respectively. The off-diagonal elements are both equal to $\sigma_{XY} = \rho_{XY}\sigma_X\sigma_Y$ and are the covariance between X and Y. The covariance matrix is symmetric because $\text{Cov}(X, Y) = \text{Cov}(Y, X)$. Similar to the univariate case in that the variance of a random variable must be positive, the covariance matrix $\boldsymbol{\Sigma}$ is positive definite.

When the distribution of (X, Y) is bivariate normal with a density (8.21), the marginal distributions of X and Y can be obtained by using (8.7). The resulting marginal distributions $f_X(x)$ and $f_Y(y)$ are of the form (7.12) and hence, both are univariate normally distributed. That is, if

$$\begin{pmatrix} X \\ Y \end{pmatrix} \sim N\left[\boldsymbol{\mu} = \begin{pmatrix} \mu_X \\ \mu_Y \end{pmatrix}, \boldsymbol{\Sigma} = \begin{pmatrix} \sigma_X^2 & \sigma_{XY} \\ \sigma_{XY} & \sigma_Y^2 \end{pmatrix}\right],$$

then

$$X \sim N(\mu_X, \sigma_X^2) \quad \text{and} \quad Y \sim N(\mu_Y, \sigma_Y^2).$$

Moreover, using (8.8), the conditional distribution of Y, given $X = x$, can be derived as and is given by

$$Y|x \sim N\left[\mu_{Y|x} = \mu_Y + \rho_{XY}\frac{\sigma_Y}{\sigma_X}(x - \mu_X), \sigma_{Y|x}^2 = \sigma_Y^2(1 - \rho_{XY}^2)\right]. \tag{8.23}$$

Specifically, the conditional distribution possesses the following properties:

- it is normal;
- its mean is a linear function of x, given by $\text{E}(Y|X = x) = \mu_{Y|x} = \mu_Y + \rho_{XY}\frac{\sigma_Y}{\sigma_X}(x - \mu_X)$;
- its variance is given by $\sigma_{Y|x}^2 = \sigma_Y^2(1 - \rho_{XY}^2)$, which is a constant and does not depend on the given x value.

8.6 Independence

The meaning of independence arose when we introduced random samples in Section 6.2.1. Having established the concepts of joint, marginal and conditional distributions in Section 8.1, a more formal definition of independence can now be presented.

Two random variables X and Y are *independent* if and only if the joint probability function equals the product of the marginal probability functions; that is, if and only if

$$f(x,y) = f_X(x)f_Y(y) = f(x)f(y). \tag{8.24}$$

Because $f(x,y) = f(x)f(y|x) = f(y)f(x|y)$ (see (8.4) and (8.5)), X and Y are independent if and only if

$$f(x|y) = f(x) \quad \text{and} \quad f(y|x) = f(y), \tag{8.25}$$

namely the conditional distribution is the same as the marginal distribution. It means that the knowledge on X will provide no additional information or will have no effect on the distribution of Y, and vice versa.

Example 8.6.1 The joint and marginal probabilities given in Table 8.2 show that $f(x,y) \neq f_X(x)f_Y(y)$, and hence the two random variables X = *household size* and Y = *number of male heads* are not independent. For example, with the knowledge that household size $X = 1$, we know that the probability of $Y = 2$ male heads is zero. Without the knowledge that $X = 1$, the probability of $Y = 2$ is 0.28. In this case, the knowledge of X provides additional information about the distribution of Y.

Example 8.6.2 Suppose that a fair dice is rolled twice and X_i is the face value of the dice in the i-th roll, $i = 1, 2$. For both X_1 and X_2, the probability of $X_i = j$ is $1/6$ for all $j = 1, \ldots, 6$. Even though the face value of the dice in the first roll is known, the probability of $X_2 = j$ remains to be $1/6$ for all j. In this case, $\Pr(X_2 = j|X_1 = x_1) = \Pr(X_2 = j)$, and X_1 and X_2 are independent.

Moreover, it can be shown that if two variables X and Y are independent, then the covariance or the correlation between X and Y is zero (Exercise 8.6.3). However, zero covariance or correlation between the two variables does not imply that the two variables are independent.

Exercise 8.6.3 Use (8.9) or (8.10) together with (8.18) and (8.24) to show that if X and Y are independent, then their covariance is zero.

8.7 Multivariate distribution

The concepts that have been introduced in the context of bivariate distributions can be generalized to multivariate distributions. For example, let $\mathbf{X}$ =

$(X_1, \ldots, X_p)^T$ be a vector that stores the p random variables with *joint probability function*

$$f(\mathbf{x}) = f(x_1, \ldots, x_p) \tag{8.26}$$

where $\mathbf{x} = (x_1, \ldots, x_p)^T$. The function $f(\mathbf{x})$ possesses properties similar to those in either (8.1) or (8.6), depending on whether the variables are discrete or continuous. Generalizing (8.2), the *marginal probability function* for X_k can be obtained by summing (for discrete random variables) or integrating (for continuous random variables) the joint probability function with respect to all other $x_i's, i = 1, \ldots, p, i \neq k$; that is,

$$f_k(x_k) = \sum_{x_1} \cdots \sum_{x_{k-1}} \sum_{x_{k+1}} \cdots \sum_{x_p} f(x_1, \ldots, x_p), \quad \text{or} \tag{8.27}$$

$$f_k(x_k) = \int_{x_1} \cdots \int_{x_{k-1}} \int_{x_{k+1}} \cdots \int_{x_p} f(x_1, \ldots, x_p) dx_p \ldots dx_{k+1} dx_{k-1} \ldots dx_1. \tag{8.28}$$

The *joint marginal distribution* of several variables can be obtained in a similar manner. Without loss of generality, let $\mathbf{X}$ be partitioned as $\mathbf{X}_1$ and $\mathbf{X}_2$, and $\mathbf{x}$ as $\mathbf{x}_1$ and $\mathbf{x}_2$, where $\mathbf{X}_1$ and $\mathbf{x}_1$ are of dimension p_1, and $\mathbf{X}_2$ and $\mathbf{x}_2$ are of dimension p_2, and $p_1 + p_2 = p$, then the joint marginal distribution $f(\mathbf{x}_2)$ for $\mathbf{X}_2$ can be obtained similar in (8.27) or (8.28) by summing or integrating over all the x_i's in $\mathbf{x}_1$.

Generalizing (8.4), the conditional joint distribution of $\mathbf{X}_1$, given $\mathbf{X}_2 = \mathbf{x}_2$, is

$$f(\mathbf{x}_1 | \mathbf{x}_2) = \frac{f(\mathbf{x}_1, \mathbf{x}_2)}{f(\mathbf{x}_2)}. \tag{8.29}$$

Moreover, generalizing (8.24), the random variables $X_1, X_2, \ldots, X_p$ are mutually independent if and only if the joint probability function equals the product of the marginal probability functions; that is, if and only if

$$f(\mathbf{x}) = f(x_1, \ldots, x_p) = f_1(x_1) f_2(x_2) \cdots f_p(x_p). \tag{8.30}$$

For each of the X_i in $\mathbf{X}$, we can find the expected value $\mathrm{E}(X_i)$ and the variance $\mathrm{Var}(X_i)$. For each pair of variables X_i and X_j we can use (8.16), (8.17) or (8.18) to find their covariance $\mathrm{Cov}(X_i, X_j)$. In a multivariate analysis, it is convenient to express all these quantities in a vector or in a matrix as follows:

$$\mathrm{E}(\mathbf{X}) = \boldsymbol{\mu} = \begin{pmatrix} \mathrm{E}(X_1) \\ \mathrm{E}(X_2) \\ \vdots \\ \mathrm{E}(X_p) \end{pmatrix}, \tag{8.31}$$

$$\text{Var}(\mathbf{X}) = \text{Cov}(\mathbf{X}) = \mathbf{\Sigma} = \begin{pmatrix} \text{Cov}(X_1, X_1) & \text{Cov}(X_1, X_2) & \dots & \text{Cov}(X_1, X_p) \\ \text{Cov}(X_2, X_1) & \text{Cov}(X_2, X_2) & \dots & \text{Cov}(X_2, X_p) \\ \vdots & \vdots & & \vdots \\ \text{Cov}(X_p, X_1) & \text{Cov}(X_p, X_2) & \dots & \text{Cov}(X_p, X_p) \end{pmatrix}$$
$$= \begin{pmatrix} \sigma_{11} & \sigma_{12} & \dots & \sigma_{1p} \\ \sigma_{21} & \sigma_{22} & \dots & \sigma_{2p} \\ \vdots & \vdots & & \vdots \\ \sigma_{p1} & \sigma_{p2} & \dots & \sigma_{pp} \end{pmatrix}. \tag{8.32}$$

The vector in (8.31) is called the mean vector and is denoted by $\text{E}(\mathbf{X})$ while the matrix in (8.32) is called the covariance matrix and is usually denoted by $\text{Var}(\mathbf{X})$, $\text{Cov}(\mathbf{X})$ or $\mathbf{\Sigma}$. Because the covariance of a variable with itself is the variance of the variable (see (8.19)), the diagonal elements $\{\sigma_{ii}\}$ in $\mathbf{\Sigma}$ are the variances of the variables. Let $\mathbf{D}_\sigma$ be a $p \times p$ diagonal matrix with $\{\sigma_{ii}\}$ as its diagonal elements and $\mathbf{D}_\sigma^{-1/2}$ be a $p \times p$ diagonal matrix with diagonal elements $\{\sigma_{ii}^{-1/2} = \sigma_i^{-1} = 1/\sqrt{\sigma_{ii}}\}$, then the matrix

$$\mathbf{\Omega} = \mathbf{D}_\sigma^{-1/2} \mathbf{\Sigma} \mathbf{D}_\sigma^{-1/2} \tag{8.33}$$

is called a *correlation matrix*. All the diagonal elements of a correlation matrix are equal to 1 and its (i, j)-th diagonal element are equal to the correlation of X_i and X_j, given by

$$\rho_{ij} = \frac{\sigma_{ij}}{\sigma_i \sigma_j}. \tag{8.34}$$

The sample quantities $\overline{X}_i$, S_i^2, and r_{ij} are estimates of the corresponding population parameters μ_i, $\sigma_i^2 = \sigma_{ii}$ and ρ_{ij}. In the context of multivariate analysis, these sample quantities can also be organized in a sample mean vector $\overline{\mathbf{X}}$ and a sample covariance matrix $\mathbf{S}$ in a way similar to their population analogues in (8.31) and (8.32).

Using some of the basic properties of mathematical expectations, such as those described in Section 7.4, it can be shown that if $\mathbf{A}$, $\mathbf{B}$ and $\mathbf{C}$ are matrices in which all of the elements are constants rather than random variables, then

$$\text{E}(\mathbf{AXB} + \mathbf{C}) = \mathbf{A}\text{E}(\mathbf{X})\mathbf{B} + \mathbf{C}, \tag{8.35}$$

$$\text{Cov}(\mathbf{AX}) = \mathbf{A}\text{Cov}(\mathbf{X})\mathbf{A}^T. \tag{8.36}$$

In (7.18), we introduced the quantity $(x - \mu)/\sigma$ or $(x_i - \bar{x})/s$, which is a distance measure using standard deviation as the unit of measurement. The multivariate analogue of such a distance measure can be similarly constructed as

$$d(\mathbf{x}, \boldsymbol{\mu}) = (\mathbf{x} - \boldsymbol{\mu})^T \mathbf{\Sigma}^{-1} (\mathbf{x} - \boldsymbol{\mu}) \tag{8.37}$$

at the population level, or

$$d(\mathbf{x}, \bar{\mathbf{x}}) = (\mathbf{x} - \bar{\mathbf{x}})^T \mathbf{S}^{-1} (\mathbf{x} - \bar{\mathbf{x}}) \tag{8.38}$$

at the sample level. This distance is called the *Mahalanobis distance.*

Exercise 8.7.1 Prove (8.35) and (8.36).

Chapter 9

Simple Linear Regression

Linear regression is a popular and widely used statistical method. It involves establishing a statistical model that depicts the relationship between a response variable and one or more independent or predictor variables, the estimating of the model parameters and the subsequent use of the model to draw statistical inferences, such as predicting the value of the response variable for a given value of the predictor variable. Model diagnoses and influence analyses that involve the identification of problematic areas in the analyzing process and observations exerting undue influence on such analyses are also important. In this chapter, we consider a *simple linear regression* with one independent variable. The generalization to a *multiple regression* with more independent variables is presented in the next chapter.

9.1 The model

Suppose we have a set of data on two variables that are represented by $\{(X_i, Y_i), i = 1, \ldots, n\}$. If the *scatter plot* of these data points in the $xy-$plane suggests that the relationship between X and Y can be reasonably described by a straight line, then it is sensible to conduct a simple linear regression analysis.

Simple linear regression assumes a linear relationship between X and Y that is not exact and is subject to random errors. The *independent variable*, also called the *predictor* variable or the *regressor*, is usually labeled as the X variable. This variable is not random and is considered to be controllable by the researcher in the sense that its values are known or fixed in advance. The variable Y, however, is considered to be the random response variable. There is also a distribution for the response variable Y for any given X value. The mean of the distribution depends on the given X value, and hence is a function of X. Moreover, it is usually assumed that the variances of the distributions are the same for the different distributions of Y with different X values. Figure

9.1 depicts the linear regression model.

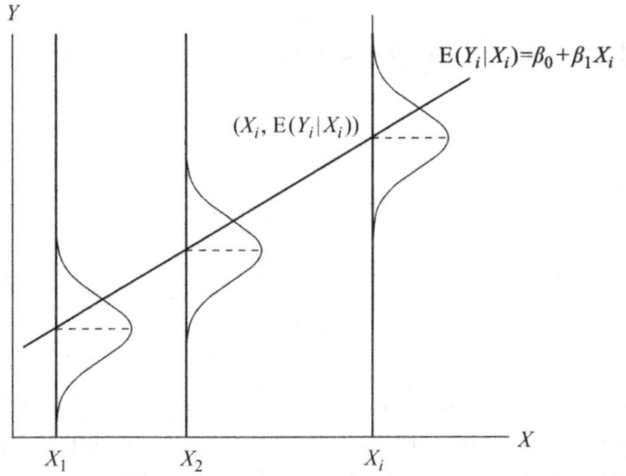

Fig. 9.1 The Simple linear regression model

Mathematically, the linear regression model is expressed as

$$Y_i = \beta_0 + \beta_1 X_i + \epsilon_i \qquad \text{for} \quad i = 1, \ldots, n, \tag{9.1}$$

where X_i represents the independent variable, Y_i is the response variable and ϵ_i is the random error, which is a random variable with a mean of 0 and a variance σ_i^2. With the assumption of common variance, we have $\sigma_i^2 = \sigma^2$ for all i. Because ϵ_i is random, Y_i is also random. The model implies that the conditional distribution of Y_i for given X_i has a mean of $\mathrm{E}(Y_i|X_i) = \beta_0 + \beta_1 X_i$ and a variance $\mathrm{Var}(Y_i|X_i) = \mathrm{Var}(\epsilon_i) = \sigma^2$. The parameters β_0 and β_1 are the intercept and the slope of the straight line that describes the linear relationship between the means of the distributions of Y and the values of the independent variable X.

The model depicts a probabilistic rather than an exact deterministic relationship that assumes a relationship between the *means* of Y and X, but not Y and X, such that specific data points (X_i, Y_i) will not necessarily lie on the straight line. In effect, they usually do not. Y_i is regarded as a randomly observed value from the distribution of Y when X equals X_i. The mean of the distribution of Y_i is $\beta_0 + \beta_1 X_i$, and a randomly drawn Y_i from the distribution may not equal the mean of the distribution.

9.2 The least squares estimation

Based on a set of data $\{(X_i, Y_i), i = 1, \ldots, n\}$ with n pairs of observations, the first task in a regression analysis is to find the estimates $\widehat{\beta_0}$ of the intercept β_0 and $\widehat{\beta_1}$ for the slope β_1 of the straight line. The *least squares* principle is employed to produce the estimates.

For the least squares principle, recall that a straight line on the xy-plane is determined by its intercept β_0 and slope β_1. Considering a value X_i, a point on the straight line is given by $(X_i, \beta_0 + \beta_1 X_i)$. The deviation of the observed Y_i value from the straight line, measured by the vertical distance, is $Y_i - (\beta_0 + \beta_1 X_i)$, as shown in Figure 9.2. The sum of squares of the deviations for all of the data points is

$$d(\beta_0, \beta_1) = \sum_{i=1}^{n} (Y_i - (\beta_0 + \beta_1 X_i))^2. \tag{9.2}$$

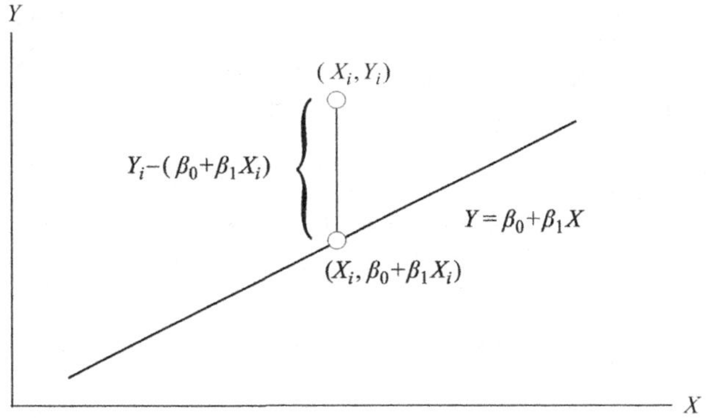

Fig. 9.2 The deviation

Different sets of β_0 and β_1 values produce different straight lines with different d values. A straight line that produces a d value that is as small as possible is desirable because the line is considered "closest" to the observed data points. Such a straight line is used as an estimate of the line in Figure 9.1, which depicts the relationship between X and the mean of Y. The line can be obtained by minimizing (9.2) with respect to β_0 and β_1 and is called the *least squares regression line.*

Let $\widehat{\beta_0}$ and $\widehat{\beta_1}$ be the solutions of β_0 and β_1, respectively, for the following equations:

$$\begin{aligned} \frac{\partial d}{\partial \beta_0} &= \frac{\partial}{\partial \beta_0} \sum_{i=1}^{n} (Y_i - (\beta_0 + \beta_1 X_i))^2 \\ &= -2\sum_{i=1}^{n} Y_i + 2n\beta_0 + 2\beta_1 \sum_{i=1}^{n} X_i = 0, \end{aligned} \tag{9.3}$$

$$\frac{\partial d}{\partial \beta_1} = \frac{\partial}{\partial \beta_1} \sum_{i=1}^{n} (Y_i - (\beta_0 + \beta_1 X_i))^2$$
$$= -2 \sum_{i=1}^{n} (Y_i - (\beta_0 + \beta_1 X_i)) X_i = 0. \tag{9.4}$$

It is now immediate that the first equation that yields

$$\widehat{\beta_0} = \overline{Y} - \widehat{\beta_1} \overline{X}, \tag{9.5}$$

where

$$\overline{X} = \frac{1}{n} \sum_{i=1}^{n} X_i \quad \text{and} \quad \overline{Y} = \frac{1}{n} \sum_{i=1}^{n} Y_i$$

are the means of X and Y, respectively. Using these results together with (9.4), (7.8) and (8.15), one obtains

$$\widehat{\beta_1} = \frac{\sum_{i=1}^{n} (X_i - \overline{X})(Y_i - \overline{Y})}{\sum_{i=1}^{n} (X_i - \overline{X})^2} = \frac{\sum_{i=1}^{n} (X_i - \overline{X}) Y_i}{\sum_{i=1}^{n} (X_i - \overline{X})^2} = \frac{SS_{XY}}{SS_{XX}}, \tag{9.6}$$

where

$$SS_{XX} = \sum_{i=1}^{n} (X_i - \overline{X})^2 \tag{9.7}$$

and

$$SS_{XY} = \sum_{i=1}^{n} (X_i - \overline{X})(Y_i - \overline{Y}) = \sum_{i=1}^{n} (X_i - \overline{X}) Y_i. \tag{9.8}$$

Similarly, we define

$$SS_{YY} = \sum_{i=1}^{n} (Y_i - \overline{Y})^2. \tag{9.9}$$

The terms SS_{XX} and SS_{YY} are called the *sum of squared deviations from the mean*, while the term SS_{XY} is called the *sum of the cross products of deviations.*

The $\widehat{\beta_0}$ and $\widehat{\beta_1}$ in (9.5) and (9.6) are the *least squares estimators* of β_0 and β_1, respectively. In fact, if we compute the Hessian matrix of the function d with respect to β_0 and β_1 and evaluate at $\beta_0 = \widehat{\beta_0}$ and $\beta_1 = \widehat{\beta_1}$, we get

$$2 \begin{pmatrix} n & n\overline{X} \\ n\overline{X} & \sum_i X_i^2 \end{pmatrix}.$$

The trace of this matrix is apparently positive and its determinant is also positive due to

$$4n \sum_{i=1}^{n} X_i^2 - 4n^2 \overline{X}^2 = 4n SS_{XX}.$$

It follows that both eigenvalues of the Hessian matrix are positive and hence the Hessian matrix is a positive-definite matrix. Therefore, the function d is minimized when $\beta_0 = \widehat{\beta_0}$ and $\beta_1 = \widehat{\beta_1}$, as explained in Section 1.2.3.

Furthermore, it can be shown that they are unbiased estimators (see Exercise 9.2.2), namely

$$\mathrm{E}(\widehat{\beta_0}) = \beta_0 \quad \text{and} \quad \mathrm{E}(\widehat{\beta_1}) = \beta_1. \tag{9.10}$$

Given a set of data, one can compute the actual numerical estimates using (9.5) and (9.6). It is worthy of note that depending on the context of the discussion, $\widehat{\beta_0}$ and $\widehat{\beta_1}$ may be used to denote the analytical expressions or the numerical values. In the former case, they are called the *least squares estimators* and in the latter case they are called the *least squares estimates.*

$\widehat{\beta_0}$ and $\widehat{\beta_1}$ simultaneously determine the *least squares regression line*

$$\widehat{Y} = \widehat{\beta_0} + \widehat{\beta_1} X. \tag{9.11}$$

This is also called the *least squares line*, the *regression line* or the *least squares prediction line*, and it represents the best to the data $\{(X_i, Y_i), i = 1, \ldots, n\}$ according to the principle of least squares.

Example 9.2.1 Table 9.1 illustrates the computation for finding the least squares estimates. For the data set $(X_i, Y_i) = \{(1,1), (3,1), (5,2), (7,2), (9,4)\}$ with 5 data points, the following results are obtained:

$$\overline{X} = \frac{25}{5} = 5, \quad \overline{Y} = \frac{10}{5} = 2,$$

$$\widehat{\beta_1} = \frac{SS_{XY}}{SS_{XX}} = \frac{14}{40} = 0.35, \quad \widehat{\beta_0} = \overline{Y} - \widehat{\beta_1}\overline{X} = 2 - 0.35 \times 5 = 0.25,$$

$$\widehat{Y} = \widehat{\beta_0} + \widehat{\beta_1} X = 0.25 + 0.35X.$$

In the table, we have also shown the computation of other quantities, such as SSE, which are be introduced later in this chapter.

Table 9.1 Least squares estimation

X_i	Y_i	$X_i - \overline{X}$	$Y_i - \overline{Y}$	$(X_i - \overline{X})^2$	$(Y_i - \overline{Y})^2$	$(X_i - \overline{X}) \times (Y_i - \overline{Y})$	$\widehat{Y_i}$	$Y_i - \widehat{Y_i}$	$(Y_i - \widehat{Y_i})^2$
1	1	–4	–1	16	1	4	0.6	0.4	0.16
3	1	–2	–1	4	1	2	1.3	–0.3	0.09
5	2	0	0	0	0	0	2	0	0
7	2	2	0	4	0	0	2.7	–0.7	0.49
9	4	4	2	16	4	8	3.4	0.6	0.36
25	10	0	0	40	6	14	10	0	1.1
$\overline{X} = 5$	$\overline{Y} = 2$			(SS_{XX})	(SS_{YY})	(SS_{XY})			(SSE)

Exercise 9.2.2 Using the basic properties of expectation described in Section 7.4, the expressions of $\widehat{\beta_0}$ and $\widehat{\beta_1}$ in (9.5) and (9.6) and the fact that $X_i's$ are given constants that are not random, show that $E(\widehat{\beta_0}) = \beta_0$ and $E(\widehat{\beta_1}) = \beta_1$.

9.3 The maximum likelihood estimation of regression parameters

When the least squares method is used to find the estimators of the regression parameters, there is no need to impose any distributional assumption on the data. However, if it is sensible to assume that ϵ_i, and hence Y_i, follows the normal distribution, then we can also find the estimators of β_0 and β_1 using the maximum likelihood (ML) method. We will see that the resulting ML estimators are the same as the least squares estimators.

The normal assumption suggests that if the distribution of ϵ_i is $N[0, \sigma^2]$, then the distribution of Y_i, given X_i, is $N[\beta_0 + \beta_1 X_i, \sigma^2]$. For a set of independently observed data $\{(X_i, Y_i), i = 1, \ldots, n\}$, the likelihood function is given by (see (7.12) and (6.9)),

$$\begin{aligned} L^*(\beta_0, \beta_1, \sigma^2) &= \prod_{i=1}^{n} \frac{1}{\sqrt{2\pi\sigma^2}} \exp\left(-\frac{(Y_i - (\beta_0 + \beta_1 X_i))^2}{2\sigma^2}\right) \\ &= (2\pi\sigma^2)^{-n/2} \exp\left(-\frac{\sum_{i=1}^{n}(Y_i - (\beta_0 + \beta_1 X_i))^2}{2\sigma^2}\right). \end{aligned}$$

Maximizing $L^*(\beta_0, \beta_1, \sigma^2)$ is equivalent to minimizing the negative log-likelihood

$$\begin{aligned} L(\beta_0, \beta_1, \sigma^2) &= -\ln L^*(\beta_0, \beta_1, \sigma^2) \\ &= \frac{n}{2} \ln(2\pi\sigma^2) + \frac{\sum_{i=1}^{n}(Y_i - (\beta_0 + \beta_1 X_i))^2}{2\sigma^2}. \end{aligned} \tag{9.12}$$

As a result, the ML estimators $\widehat{\beta_0}_{ML}$ of β_0 and $\widehat{\beta_1}_{ML}$ of β_1 are obtained by minimizing the quantity

$$\sum_{i=1}^{n}(Y_i - (\beta_0 + \beta_1 X_i))^2$$

which is the same as $d(\beta_0, \beta_1)$ in (9.2). In other words, the ML estimators $\widehat{\beta_0}_{ML}$ of β_0 and $\widehat{\beta_1}_{ML}$ of β_1 are the same as the least square estimators $\widehat{\beta_0}$ and $\widehat{\beta_1}$

in (9.5) and (9.6). Moreover, to find the ML estimator $\widehat{\sigma}^2_{ML}$ of σ^2, we take the partial derivative for the negative log-likelihood in (9.12) with respect to σ^2 and set the derivative to zero. The solution of σ^2 for the likelihood equation

$$\frac{\partial L(\beta_0, \beta_1, \sigma^2)}{\partial \sigma^2} = \frac{n}{2\sigma^2} - \frac{\sum_{i=1}^{n}(Y_i - (\beta_0 + \beta_1 X_i))^2}{2(\sigma^2)^2} = 0$$

with β_0 and β_1 replaced by their ML estimators $\widehat{\beta_0}_{ML} = \widehat{\beta_0}$ and $\widehat{\beta_1}_{ML} = \widehat{\beta_1}$ gives the ML estimator of σ^2. That is,

$$\widehat{\sigma}^2_{\text{ML}} = \frac{1}{n}\sum_{i=1}^{n}(Y_i - (\widehat{\beta_0} + \widehat{\beta_1}X_i))^2 = \frac{SSE}{n}, \tag{9.13}$$

where $SSE = \sum_{i=1}^{n}(Y_i - (\widehat{\beta_0} + \widehat{\beta_1}X_i))^2$ is the sum of the squared *residuals* discussed in the next section.

9.4 Residuals

For each data point, the value

$$r_i = Y_i - \widehat{Y_i} = Y_i - (\widehat{\beta_0} + \widehat{\beta_1}X_i) \tag{9.14}$$

is called the *residual*. It is the deviation of the observed value Y_i from the fitted value $\widehat{Y_i}$ which is determined by the regression line (9.11). From (9.1) or Figure 9.1, $\widehat{Y_i}$ can be regarded as an estimate of the mean of the distribution of Y when $X = X_i$.

For the data set in Table 9.1, the residuals for the five data points are respectively, 0.4, −0.3, 0, −0.7 and 0.6. The sum of the residuals is zero, which is true for any data set because according to (9.11), (9.5) and (6.5),

$$\begin{aligned}
\sum_{i=1}^{n} r_i &= \sum_{i=1}^{n}(Y_i - \widehat{Y_i}) \\
&= \sum_{i=1}^{n}(Y_i - (\widehat{\beta_0} + \widehat{\beta_1}X_i)) \\
&= \sum_{i=1}^{n}(Y_i - (\overline{Y} - \widehat{\beta_1}\overline{X} + \widehat{\beta_1}X_i)) \\
&= \sum_{i=1}^{n}(Y_i - \overline{Y}) - \widehat{\beta_1}\sum_{i=1}^{n}(X_i - \overline{X}) \\
&= 0.
\end{aligned} \tag{9.15}$$

Summing over all the data points their squared residuals gives

$$SSE = \sum_{i=1}^{n} r_i^2 = \sum_{i=1}^{n} (Y_i - (\widehat{\beta_0} + \widehat{\beta_1} X_i))^2. \tag{9.16}$$

This is called the *residual sum of squares*, also known as the *sum of squares due to error*. It is the minimum value for d in (9.2). For example, the SSE for the data in Table 9.1 is given by

$$\sum_{i=1}^{n} r_i^2 = \sum_{i=1}^{n} (Y_i - \widehat{Y_i})^2 = (1 - 0.6)^2 + \cdots + (4 - 3.4)^2 = 0.4^2 + \cdots + 0.6^2 = 1.1.$$

Each residual is the difference between the observed Y value and the estimate of the mean of its distribution. It therefore provides information on σ^2 that represents the dispersion of the Y distributions. The residual sum of squares SSE aggregates the squared residuals. The average, namely $SSE/n = \sum_{i=1}^{n} r_i^2/n$, is therefore a sensible and intuitively appealing estimate of σ^2. In effect, we show in (9.13) that $\widehat{\sigma}_{ML}^2 = SSE/n$ is the ML estimator of σ^2 if ϵ_i, or equivalently $Y_i, i = 1, \ldots, n$, follow a normal distribution with the common variance σ^2. However, in a regression analysis, it is more common to estimate σ^2 by the *mean squared error*

$$\widehat{\sigma^2} = MSE = \frac{SSE}{n-2} = \frac{\sum_{i=1}^{n} r_i^2}{n-2} = \frac{\sum_{i=1}^{n} (Y_i - (\widehat{\beta_0} + \widehat{\beta_1} X_i))^2}{n-2}. \tag{9.17}$$

For example, for the data set in Table 9.1, we have

$$\widehat{\sigma^2} = MSE = \frac{SSE}{n-2} = \frac{1.1}{3} = 0.3667. \tag{9.18}$$

Similar to $\widehat{\beta_0}$ and $\widehat{\beta_1}$, the $\widehat{\sigma^2}$ in (9.17) with an analytical expression is called *an estimator* and the $\widehat{\sigma^2}$ in (9.18) with a numerical value is called *an estimate*, although both are denoted by the same notation.

The MSE in (9.17) is a preferable estimator for σ^2 because it can be shown as unbiased in the sense that

$$\mathrm{E}(MSE) = \mathrm{E}\left[\frac{\sum_{i=1}^{n} (Y_i - (\widehat{\beta_0} + \widehat{\beta_1} X_i))^2}{n-2}\right] = \sigma^2. \tag{9.19}$$

Nevertheless, the difference between n and $n-2$ is negligible when sample size n is sufficiently large. Specifically, the maximum likelihood estimator $\widehat{\sigma}_{ML}^2 = SSE/n$ is also unbiased when the sample size is large, or it is *asymptotically unbiased*.

9.5 Coefficient of determination

Given a set of data on two variables, one can always fit a regression line. Whether or not the resulting regression line is sensible and can adequately describe the relationship between the two variables depends on whether the strength of the linear relationship between the two variables is strong. Atthough we can use (8.13) to compute the correlation coefficient and examine the strength of the linear relationship, other strength measure, that can be easily generalized to a multiple regression that involves more than one independent variable is the *coefficient of determination.*

The coefficient of determination is constructed by using the components that are obtained by partitioning the total sum of squares of the dependent variable Y. It assesses the proportion of variation of Y that can be explained by the regression model or, equivalently, the proportion of variation that can be attributed to the independent variable X. Considering the set of n-many Y values, a measure of their total variation is given by

$$SST = SS_{YY} = \sum_{i=1}^{n}(Y_i - \overline{Y})^2. \tag{9.20}$$

This total variation is the aggregation of the quantity $Y_i - \overline{Y}$. An illustration based on the data points in Table 9.1 is given in the upper chart of Figure 9.3. In a regression analysis, each measurement of Y is accompanied by a measurement of X, such that the data points can be plotted in an xy-plane as shown in the lower chart in Figure 9.3. Attempts at a regression analysis are to see whether we can use the linear relationship between X and the mean of Y to explain the variation of Y. The value of the y-coordinate for a point on the regression line can be regarded as the estimated mean of the distribution of Y for a given X value and can also be used to predict a value of Y for a given X value. For example, if the slope of the regression line is positive, we would anticipate that a large Y value is likely to come with a large X value, and hence X can explain part of the variation of Y. Specifically, let

$$Y_i - \overline{Y} = (Y_i - \widehat{Y}_i) + (\widehat{Y}_i - \overline{Y}),$$

then we can decompose the deviation of Y_i from its mean $\overline{Y}$ into two components, which is shown in Figure 9.3(b). The component $\widehat{Y}_i - \overline{Y}$ represents the deviation of Y_i from the mean of Y, which is predicted by the regression model or that can be attributed to the effect of X. It is the variation of Y that can be explained by the regression model. The other component is $Y_i - \widehat{Y}_i$,which is the amount of deviation that remains unexplained after we have taken the effect of the variable X into consideration. It is also the amount of variation due to random error.

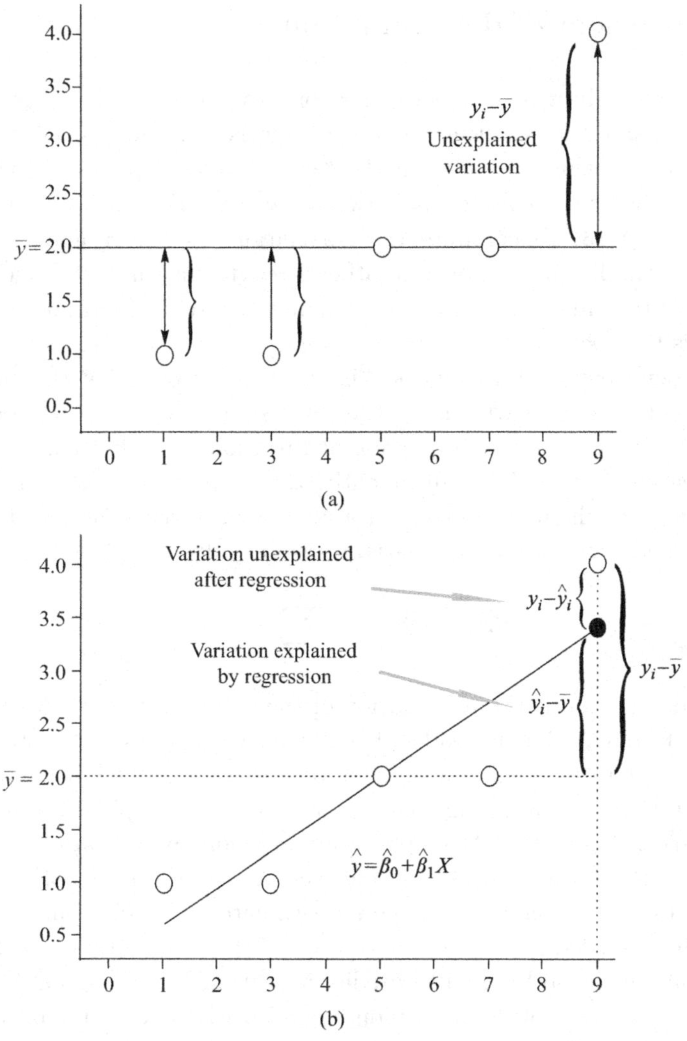

Fig. 9.3 Partitioning of total variance

Given facts that $\sum_{i=1}^{n} r_i = 0$ (see (9.15)) and $\sum_{i=1}^{n} X_i r_i = 0$ (see Exercise 9.5.2), it can be shown that $\sum_{i=1}^{n}(\widehat{Y}_i - \overline{Y})(Y_i - \widehat{Y}_i) = 0$. As a result, the total variation of Y can be partitioned into two components as follows:

$$\begin{aligned} SST &= \sum_{i=1}^{n}(Y_i - \overline{Y})^2 \\ &= \sum_{i=1}^{n}(\widehat{Y}_i - \overline{Y})^2 + \sum_{i=1}^{n}(Y_i - \widehat{Y}_i)^2 \\ &= SSR + SSE. \end{aligned}$$

The component $SSR = \sum_{i=1}^{n}(\widehat{Y}_i - \overline{Y})^2$ is called the *variation due to regression* and is the variation that can be attributed to the relationship between X and Y. $SSE = \sum_{i=1}^{n}(Y_i - \widehat{Y}_i)^2$ is the *residual sum of squares*, which is the variation due to random errors. The proportion of variation that can be explained by the regression model is given by

$$R^2 = \frac{SSR}{SST} = \frac{SST - SSE}{SST} = 1 - \frac{SSE}{SST},$$

and R^2 is called the *coefficient of determination* . The positive square root of R^2, namely $R = \sqrt{R^2}$, is called the *multiple correlation coefficient* or the *coefficient of multiple correlation.*

Because $SST = SSR + SSE$, it is obvious that $0 \leqslant R^2 \leqslant 1$ and, equivalently, $0 \leqslant R \leqslant 1$. If all of the data points lie on the regression line, then $SSE = 0$ and $R^2 = R = 1$. In this case, the simple linear regression provides a perfect fit to the data points. The larger the coefficient is, the larger the proportion of variation that can be explained by the regression model, which suggests a better fit to the regression model.

In simple linear regression with only one independent variable, it is clear that (see Exercise 9.5.1) through direct substitution, the coefficient of multiple determination R^2 is the same as for r^2, where r is the sample correlation coefficient in (8.13). While r is mainly used to measure the association between two variables, R is easily generalized to a multiple regression model in which there is more than one independent variable.

Exercise 9.5.1 Show that $R^2 = \frac{SS_{XY}^2}{SS_{XX}SS_{YY}} = r^2$, where $SS_{XX}, SS_{XY}, SS_{YY}$ and r are defined in (9.7), (9.8), (9.9) and (8.13), respectively.

Exercise 9.5.2 Use (9.4) to show that $\sum_{i=1}^{n} X_i r_i = 0$.

9.6 Weighted least squares estimates

Each data point (X_i, Y_i) is treated as the same when we apply the least squares principle or the ML method to find the estimates of β_0 and β_1 by minimizing the discrepancy function in (9.2). Frequently, we place greater certain data points, which can be achieved by using the *weighted least squares method.* Let $\omega_1, \ldots, \omega_n$ be a set of known weights that are assigned to the n data points, then the *weighted least squares estimates* of β_0 and β_1 are obtained by minimizing

$$d_\omega = \sum_{i=1}^{n} \omega_i (Y_i - (\beta_0 + \beta_1 X_i))^2 \tag{9.21}$$

with respect to β_0 and β_1. The minimum can then be achieved by

$$\widehat{\beta_{1_\omega}} = \frac{\sum \omega_i X_i Y_i - \dfrac{\sum \omega_i X_i \sum \omega_i Y_i}{\sum \omega_i}}{\sum \omega_i X_i^2 - \dfrac{(\sum \omega_i X_i)^2}{\sum \omega_i}}$$

and

$$\widehat{\beta_{0\omega}} = \frac{\sum \omega_i Y_i - \widehat{\beta_{1_\omega}} \sum \omega_i X_i}{\sum \omega_i},$$

where all summations go from $i = 1$ to n.

The weighted least squares method is commonly used when the variances of the distributions for Y with different X values are different; that is, when $\text{Var}(\epsilon_i) = \text{Var}(Y_i) = \sigma_i^2$ varies with i. In this case, those Y values from the distributions with smaller variances are more reliable and hence, it is desirable that the estimation rely more on these values. This can be achieved by assigning a larger weights to such data points. For example, a set of weights that are inversely proportional to the variance of ϵ_i, such as, $\omega_i = 1/\sigma_i^2$ for all i, can be employed in a weighted least squares estimation.

Chapter 10

Topics on Linear Regression Analysis

Chapter 9 provides an introduction to simple linear regression, but there are many topics in regression that are worth further attention. In this chapter, we address some such topics that are relevant to our subsequent discussion on local influence in Part III.

10.1 Multiple regression model

A simple linear regression model can be generalized to a *multiple regression model* for analyzing the effect of two or more independent variables on the response variable. Generalizing (9.1), the multiple regression model is given by

$$Y_i = \beta_0 + \beta_1 X_{i1} + \beta_2 X_{i2} + \cdots + \beta_k X_{ik} + \epsilon_i \qquad \text{for} \quad i = 1, \ldots, n, \tag{10.1}$$

where Y_i represents the response variable and $X_{i1}, \ldots, X_{ik}$ represent the k independent variables, which are also called predictor variables or regressors. Similar to the simple linear regression, ϵ_i is the random error with a mean of 0 and a variance of σ_i^2, and hence we have $\mathrm{E}(Y_i) = \beta_0 + \beta_1 X_{i1} + \cdots + \beta_k X_{ik}$ for given X_{i1} to X_{ik} and $\mathrm{Var}(Y_i) = \sigma_i^2$. When all of the variances are assumed to be the same, we have $\sigma_i^2 = \sigma^2$ for all i. The equation in (10.1), without the error term ϵ_i, determines a mean response plane. The parameter β_0 is the intercept of the mean response plane and gives the mean of Y when all of the X's equal zero. Each parameter β_j, $j = 1, \ldots, k$ represents the change of the mean of Y per unit change of X_j when all other $X_{j'}, j' = 1, \ldots, k, j' \neq j$ are held constant.

The multiple regression model in (10.1) is a linear model with respect to the β parameters. Because an X variable is considered non-random and its value is known, any functions of X variables can be easily constructed and be included as independent variables in the multiple regression model, such that model remains linear to the β parameters. For example, the *polynomial model with degree k* given by

$$Y_i = \beta_0 + \beta_1 X_i + \beta_2 X_i^2 + \cdots + \beta_k X_i^k + \epsilon_i$$

is treated as

$$Y_i = \beta_0 + \beta_1 X_{i1} + \beta_2 X_{i2} + \cdots + \beta_k X_{ik} + \epsilon_i$$

with $X_{il} = X_i^l$ in a multiple regression model. It is also regarded as a linear model.

10.2 Estimation and interpretation

Let $\boldsymbol{\beta} = (\beta_0, \beta_1, \ldots, \beta_k)^T$. The estimates $\widehat{\beta_j}$ of β_j, for $j = 0, \ldots, k$, can be obtained, as in the simple linear regression by the least squares or maximum likelihood methods when normal assumption is tenable. Both methods reduce to minimizing the function

$$d = \sum_{i=1}^{n} (Y_i - (\beta_0 + \beta_1 X_{i1} + \cdots + \beta_k X_{ik}))^2 \tag{10.2}$$

with respect to $\beta_j, j = 0, \ldots, k$. For multiple regression, it is more convenient to make use of the matrix approach to represent the model. Let

$$\mathbf{Y} = \begin{pmatrix} Y_1 \\ \vdots \\ Y_n \end{pmatrix}, \mathbf{X} = \begin{pmatrix} 1 & X_{11} & \ldots & X_{1k} \\ 1 & X_{21} & \ldots & X_{2k} \\ \vdots & \vdots & & \vdots \\ 1 & X_{n1} & \ldots & X_{nk} \end{pmatrix}, \boldsymbol{\beta} = \begin{pmatrix} \beta_0 \\ \beta_1 \\ \vdots \\ \beta_k \end{pmatrix} \text{ and } \boldsymbol{\epsilon} = \begin{pmatrix} \epsilon_1 \\ \vdots \\ \epsilon_n \end{pmatrix},$$

the model in (10.1) can then be expressed as

$$\mathbf{Y} = \mathbf{X}\boldsymbol{\beta} + \boldsymbol{\epsilon},$$

where

$$\text{Cov}(\mathbf{Y}) = \text{Cov}(\boldsymbol{\epsilon}) = \sigma^2 \mathbf{I} \tag{10.3}$$

Based on the assumptions that $\text{Var}(Y_i) = \sigma_i^2 = \sigma^2$ for all i and that Y_i and Y_j are independent for $i \neq j$ (see Exercise 8.6.3). As a result, (10.2) can be expressed as

$$d = d(\boldsymbol{\beta}) = (\mathbf{Y} - \mathbf{X}\boldsymbol{\beta})^T (\mathbf{Y} - \mathbf{X}\boldsymbol{\beta}).$$

To find the minimum of the function $d(\boldsymbol{\beta})$, we find the zeros of its gradient and solve the set of equations

$$\frac{\partial d}{\partial \beta_j} = \frac{\partial}{\partial \beta_j} (\mathbf{Y} - \mathbf{X}\boldsymbol{\beta})^T (\mathbf{Y} - \mathbf{X}\boldsymbol{\beta}) = \mathbf{0}.$$

Because $\frac{\partial}{\partial \beta_j}\boldsymbol{\beta} = \mathbf{e}_j$ is a standard base vector,

$$0 = \frac{\partial d}{\partial \beta_j} = -2\mathbf{e}_j^T\mathbf{X}^T(\mathbf{Y} - \mathbf{X}\widehat{\boldsymbol{\beta}})$$

for each $\mathbf{e}_j$. Therefore, $\mathbf{X}^T\mathbf{Y} - \mathbf{X}^T\mathbf{X}\widehat{\boldsymbol{\beta}} = \mathbf{0}$. Furthermore,

$$\frac{\partial^2 d}{\partial \beta_j \beta_k} = 2\mathbf{e}_j^T(\mathbf{X}^T\mathbf{X})\mathbf{e}_k.$$

Thus, the Hessian of the function d is the matrix $2(\mathbf{X}^T\mathbf{X})$. Because for any vector $\mathbf{v}$, $2\mathbf{v}^T(\mathbf{X}^T\mathbf{X})\mathbf{v} = 2|\mathbf{X}\mathbf{v}|^2$, the matrix $2(\mathbf{X}^T\mathbf{X})$ is positive-definite. As noted in Section 1.2.3,

$$\widehat{\boldsymbol{\beta}} = \begin{pmatrix} \widehat{\beta_0} \\ \widehat{\beta_1} \\ \vdots \\ \widehat{\beta_k} \end{pmatrix} = (\mathbf{X}^T\mathbf{X})^{-1}\mathbf{X}^T\mathbf{Y} \tag{10.4}$$

provides a minimum for the function d and gives the least squares estimate $\widehat{\boldsymbol{\beta}}$ of $\boldsymbol{\beta}$. The set of estimates $\widehat{\beta_0}, \widehat{\beta_1}, \dots, \widehat{\beta_k}$ determines a *regression plane*. Given $X_{i1}, \dots, X_{ik}$, the point on the regression plane has

$$\widehat{Y_i} = \widehat{\beta_0} + \widehat{\beta_1}X_{i1} + \cdots + \widehat{\beta_k}X_{ik},$$

and the *residual* for the i-th data point is defined by

$$r_i = Y_i - \widehat{Y_i}. \tag{10.5}$$

Let $\widehat{\mathbf{Y}} = (\widehat{Y_1}, \dots, \widehat{Y_n})^T$ and $\mathbf{r} = (r_1, \dots, r_n)^T$, we have

$$\widehat{\mathbf{Y}} = \mathbf{X}\widehat{\boldsymbol{\beta}} = \mathbf{X}(\mathbf{X}^T\mathbf{X})^{-1}\mathbf{X}^T\mathbf{Y} = \mathbf{H}\mathbf{Y}, \tag{10.6}$$

and

$$\mathbf{r} = \mathbf{Y} - \widehat{\mathbf{Y}} = \mathbf{Y} - \mathbf{H}\mathbf{Y} = (\mathbf{I} - \mathbf{H})\mathbf{Y}, \tag{10.7}$$

where the matrix

$$\mathbf{H} = \mathbf{X}(\mathbf{X}^T\mathbf{X})^{-1}\mathbf{X}^T \tag{10.8}$$

is called the *hat matrix*. The hat matrix plays an important role in regression analysis, particularly in its influence analysis.

Recall that the matrix $\mathbf{H}$, as given above, is the orthogonal projection matrix from $\mathbb{R}^n$ on the regression plane (See Section 2.6). In particular, it is idempotent and symmetric. The function d measures the distance-squared between the vector of observed values $\mathbf{Y}$ and the vector of predicted values $\widehat{\mathbf{Y}}$

given by the regression plane. The search for $\widehat{\boldsymbol{\beta}}$ is to find a plane to minimize this quantity. The vector of residues $\mathbf{r}$ is the projection of the observed values $\mathbf{Y}$ onto the direction orthogonal to the regression plane as a consequence of Corollary 2.6.3.

Many of the results in the simple linear regression discussed in Chapter 9 can be generalized to multiple regression. Specifically, $\widehat{\boldsymbol{\beta}}$ collects the least squares estimators and is a vector of random variables. It can be shown that the least squares estimators are unbiased; that is,

$$\mathrm{E}(\widehat{\boldsymbol{\beta}}) = \boldsymbol{\beta}.$$

From $\mathrm{Cov}(\mathbf{Y}) = \sigma^2\mathbf{I}$ (see 10.3) and from the discussion in Section 8.7 and in particularly (8.36), we have

$$\begin{aligned} &\mathrm{Cov}(\widehat{\boldsymbol{\beta}}) \\ &= \mathrm{Cov}((\mathbf{X}^T\mathbf{X})^{-1}\mathbf{X}^T\mathbf{Y}) \\ &= (\mathbf{X}^T\mathbf{X})^{-1}\mathbf{X}^T\mathrm{Cov}(\mathbf{Y})\mathbf{X}(\mathbf{X}^T\mathbf{X})^{-1} = \sigma^2(\mathbf{X}^T\mathbf{X})^{-1}, \end{aligned} \tag{10.9}$$

where the variances of the regression coefficients are stored in the diagonal of $\mathrm{Cov}(\widehat{\boldsymbol{\beta}})$ (see (8.32)). We also have

$$\sum_{i=1}^{n} r_i = \mathbf{r}^T\mathbf{1} = \mathbf{Y}^T(\mathbf{I}-\mathbf{H})\mathbf{1} = 0, \tag{10.10}$$

where $\mathbf{1}$ is a vector in $\mathbb{R}^n$ and all its coordinates equal to 1. This is also the first column vector in the matrix $\mathbf{X}$. As noted in Corollary 2.6.3, while the $\mathbf{H}$ given in (10.8) is a projection on to the column space of $\mathbf{X}$, $\mathbf{I}-\mathbf{H}$ is a projection onto its orthogonal complement. Therefore, $(\mathbf{I}-\mathbf{H})\mathbf{1} = \mathbf{0}$ and the equality (10.10) holds.

The minimum of d in (10.2) is achieved when $\widehat{\boldsymbol{\beta}} = (\mathbf{X}^T\mathbf{X})^{-1}\mathbf{X}^T\mathbf{Y}$ and is the *residual sum of squares* or the *sum of squares due to error*, given by

$$SSE = \mathbf{r}^T\mathbf{r} = [(\mathbf{I}-\mathbf{H})\mathbf{Y}]^T[(\mathbf{I}-\mathbf{H})\mathbf{Y}] = \mathbf{Y}^T\mathbf{Y} - \widehat{\boldsymbol{\beta}}^T\mathbf{X}^T\mathbf{Y}. \tag{10.11}$$

The ML estimator of σ^2 is given by

$$\widehat{\sigma}^2_{ML} = \frac{SSE}{n}, \tag{10.12}$$

and an unbiased estimator of σ^2 is given by

$$\widehat{\sigma}^2 = MSE = \frac{SSE}{n-(k+1)}. \tag{10.13}$$

The discussion in Section 9.5 of simple linear regression of the coefficient of determination can also be generalized. The total variation of $\mathbf{Y}$ is

$$SST = SS_{YY} = \sum_{i=1}^{n}(Y_i - \overline{Y})^2 = |\mathbf{Y} - \overline{Y}\mathbf{1}|^2.$$

Define

$$SSR = \sum_{i=1}^{n}(\widehat{Y}_i - \overline{Y})^2 = |\widehat{\mathbf{Y}} - \overline{Y}\mathbf{1}|^2.$$

Note from (10.7) that

$$SST = |\mathbf{Y} - \widehat{\mathbf{Y}} + \widehat{\mathbf{Y}} - \overline{Y}\mathbf{1}|^2 = |(\mathbf{I} - \mathbf{H})\mathbf{Y} + \widehat{\mathbf{Y}} - \overline{Y}\mathbf{1}|^2.$$

Because $\mathbf{1}$ is in the first column of $\mathbf{X}$, according to Corollary 2.6.3 it is in the regression plane. Because $\widehat{\mathbf{Y}}$ is the projection of the vector $\mathbf{Y}$ onto the regression plan, $\widehat{\mathbf{Y}} - \overline{Y}\mathbf{1}$ is in the regression plane. However, according to Corollary 2.6.3 $(\mathbf{I} - \mathbf{H})\mathbf{Y}$ is in its orthogonal complement of the regression plane, we have a decomposition

$$SST = |(\mathbf{I} - \mathbf{H})\mathbf{Y} + \widehat{\mathbf{Y}} - \overline{Y}\mathbf{1}|^2 = |(\mathbf{I} - \mathbf{H})\mathbf{Y}|^2 + |\widehat{\mathbf{Y}} - \overline{Y}\mathbf{1}|^2 = SSE + SSR.$$

Because $\langle \mathbf{Y}, \mathbf{1}\rangle = \mathbf{Y}^T\mathbf{1} = n\overline{Y}$, $|\mathbf{1}|^2 = n$, $\mathbf{Y}^T\mathbf{H}\mathbf{1} = \mathbf{Y}^T\mathbf{1} = n\overline{Y}$ from (10.10) and $\widehat{\mathbf{Y}} = \mathbf{X}\widehat{\boldsymbol{\beta}} = \mathbf{H}\mathbf{Y}$ from (10.6), we have

$$SST = \mathbf{Y}^T\mathbf{Y} - \frac{1}{n}\mathbf{Y}^T\mathbf{1}\mathbf{1}^T\mathbf{Y}, \quad SSR = \widehat{\boldsymbol{\beta}}^T\mathbf{X}^T\mathbf{Y} - \frac{1}{n}\mathbf{Y}^T\mathbf{1}\mathbf{1}^T\mathbf{Y}.$$

The strength of the linear relationship between Y and the set of X variables can be assessed by using the *coefficient of multiple determination*

$$R^2 = \frac{SSR}{SST} = \frac{SST - SSE}{SST} = 1 - \frac{SSE}{SST},$$

or, equivalently, the *coefficient of multiple correlation* $R = \sqrt{R^2}$. It is obvious that $0 \leqslant R \leqslant 1$ and that R equals 1 if all of the Y_i's lie on the regression plane. A larger R value suggests a better fit of the model.

Although R^2 is commonly used as a convenient measure to assess the proportion of variation that can be explained by the independent variables, it should be noted that the component SSE in R^2 is the minimum of (10.2), and that the minimum value decreases with the increase of k, which is the number of X variables in the regression model. In other words, R^2 increases with k and achieves $R^2 = 1$ when k is increased to $n = k + 1$. A perfect fit model with $SSE = 0$ can always be obtained by including more independent X variables in the model, even though the X variables may be totally irrelevant to the response variable Y. This should be taken into consideration when R^2 is being interpreted, especially when n is small relative to k. Some measures that adjust for the effect of the number of variables can be found in [?] and are referenced therein.

Moreover, if we wish to place greater emphasis on specific data points, such as when the variances of ϵ_i or Y_i vary, then we may wish to assign different weights to the data points in the regression analysis. Let $\boldsymbol{\Omega}$ be a diagonal matrix with the given weights $\omega_i, i = 1, \ldots, n$ as its diagonal elements, then minimizing

$$d_{\boldsymbol{\omega}}(\boldsymbol{\beta}) = \sum_{i=1}^{n} \omega_i (Y_i - (\beta_0 + \beta_1 X_{i1} + \cdots + \beta_k X_{ik}))^2 = (\mathbf{Y} - \mathbf{X}\boldsymbol{\beta})^T \boldsymbol{\Omega} (\mathbf{Y} - \mathbf{X}\boldsymbol{\beta}) \tag{10.14}$$

with respect to $\boldsymbol{\beta}$ gives the *weighted least squares estimator*

$$\widehat{\boldsymbol{\beta}}_{\boldsymbol{\omega}} = (\mathbf{X}^T \boldsymbol{\Omega} \mathbf{X})^{-1} \mathbf{X}^T \boldsymbol{\Omega} \mathbf{Y}. \tag{10.15}$$

The derivation of the weighted least squares estimator is available in Section 10.7 (see equation (10.30)).

10.3 Influential observations and outliers

In simple linear regression or multiple regression analysis, data sets may consist of data points with extreme values that are separated from the remainder of the data. For example, Figure 10.1 illustrates several types of extreme observations in a simple linear regression analysis, three of which are well separated from the remainder of the data. Observation 1 has extreme X and extreme Y values, but its presence will not change the regression line substantially and hence, it is not a very influential data point in the regression analysis. The X value for

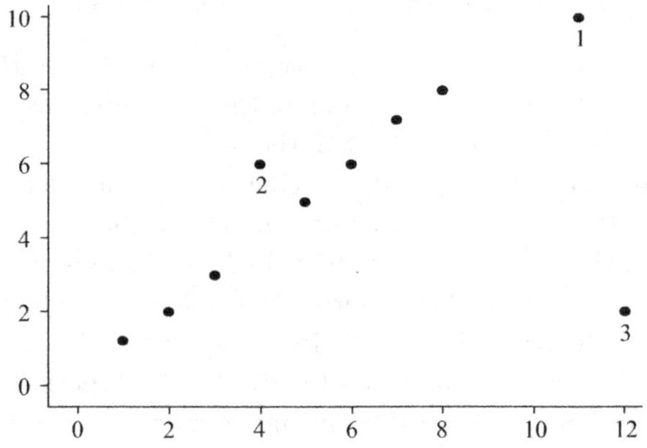

Fig. 10.1 Extreme observations in linear regression analysis

Observation 2 is close to the X values of other data points, but its Y value, when judged with reference to a linear relationship appears to be larger when compared to those of other data points. Other data points in the data set with X values close to those in Observation 2 have Y values that are consistent with the linear relation displayed by the majority of the data points. As a result, the presence of Observation 2 may not lead to a dramatic change in the regression line. However, the extreme X value in Observation 3 has undue influence on the regression line.

It is undesirable for the regression analysis result to be highly influenced by one or several data points, and hence it is important to identify data points that exert undue influence on the regression analysis. Graphical techniques are useful for the identification of extreme data points in simple linear regressions, but they become difficult to apply in multiple regressions with more independent X variables. In the next two sections, we introduce the measures *leverage* and *Cook's distance*, which are frequently used in regression analysis to identify influential observations that have undue impact.

Moreover, because the residual $r_i = Y_i - \widehat{Y}_i$ for each data point can be regarded as the observed error or an estimate of ϵ_i if the regression model has been correctly specified, various residual analysis techniques and residual plots are frequently used to flag data points that worth further attention or to detect the violation of regression model assumptions.

For example, residuals can be used to detect the violation of normal assumptions. If ϵ_i is normally distributed with a mean of 0 and a variance of σ^2, then the plot of all the residuals should exhibit a pattern of normal distribution; that is, symmetric about zero and bell-shaped. Moreover, property (7.16) for normal distribution also suggests that approximately 95.4% of the data points should have residuals that fall within $\pm 2\widehat{\sigma}$, where $\widehat{\sigma}^2$ is given by (10.13). Any data point that has a residual with an extremely large magnitude requires further attention. Specifically, data points with a residual magnitude greater than $3\widehat{\sigma}$ to $4\widehat{\sigma}$ are sometimes called *outliers* in the literature. It is worth mentioning that by adopting this definition of outliers, data points with extreme X values that are highly influential in the regression analysis may not be outliers. This point is explained further in the next section.

10.4 Leverage

Let h_{ij} be the (i, j) entry of the hat matrix $\mathbf{H} = \mathbf{X}(\mathbf{X}^T\mathbf{X})^{-1}\mathbf{X}^T$, then the diagonal element h_{ii} is called the *leverage* of the i-th data point. From (10.6), $\widehat{\mathbf{Y}} = \mathbf{X}\widehat{\boldsymbol{\beta}} = \mathbf{X}(\mathbf{X}^T\mathbf{X})^{-1}\mathbf{X}^T\mathbf{Y} = \mathbf{H}\mathbf{Y}$, hence

$$\widehat{Y}_i = h_{i1}Y_1 + h_{i2}Y_2 + \cdots + h_{ii}Y_i + \cdots + h_{in}Y_n \tag{10.16}$$

depends on the observed $Y_i, i = 1, \ldots, n$. Equation (10.16) suggests that h_{ii} can be regarded as the amount of *leverage* that Y_i has on $\widehat{Y}_i$. If h_{ii} is large, then the amount of leverage that Y_i has on the fitted $\widehat{Y}_i$ is large. In this case, the contribution of Y_i in determining $\widehat{Y}_i$ is substantial and hence, the i-th data point is a highly influential point in the regression analysis.

To further investigate the meaning of leverage, we use simple linear regression as an illustrative example. From (9.5), (9.6) and (9.8),

$$\begin{aligned}
\widehat{Y}_i &= \widehat{\beta_0} + \widehat{\beta_1} X_i \\
&= \overline{Y} + (X_i - \overline{X})\widehat{\beta_1} \\
&= \overline{Y} + (X_i - \overline{X}) \frac{\sum_{i=1}^{n} (X_i - \overline{X}) Y_i}{\sum_{i=1}^{n} (X_i - \overline{X})^2} \\
&= \cdots + \left(\frac{1}{n} + \frac{(X_i - \overline{X})^2}{\sum_{i=1}^{n} (X_i - \overline{X})^2} \right) Y_i + \cdots,
\end{aligned}$$

such that

$$h_{ii} = \frac{1}{n} + \frac{(X_i - \overline{X})^2}{\sum_{i=1}^{n} (X_i - \overline{X})^2}. \tag{10.17}$$

This shows that h_{ii} is a measure of the distance between the X_i value and the mean of all of the X values, with a large h_{ii} value resulting if the i-th observation is far away from the center of the X observations. For example, Observations 1 and 3 in Figure 10.1 are high leverage, influential points while Observation 2 is not influential. This interpretation can also be generalized to multiple regression with many X variables.

Moreover, because $\mathbf{r} = (\mathbf{I} - \mathbf{H})\mathbf{Y}$ (see (10.7)) and the hat matrix $\mathbf{H}$ involves only the non-random quantity $\mathbf{X}$, which is fixed in advance, from (10.3) and (8.36) we have

$$\mathrm{Cov}(\mathbf{r}) = (\mathbf{I} - \mathbf{H})\mathrm{Cov}(\mathbf{Y})(\mathbf{I} - \mathbf{H})^T = (\mathbf{I} - \mathbf{H})\sigma^2 \mathbf{I} (\mathbf{I} - \mathbf{H})^T = \sigma^2 (\mathbf{I} - \mathbf{H}).$$

Its i-th diagonal element gives

$$\mathrm{Var}(r_i) = \sigma^2 (1 - h_{ii}), \tag{10.18}$$

where $h_{ii} \leqslant 1$ according to Corollary 2.6.4. The result in (10.18) shows that when the leverage h_{ii} is large, the variance of r_i is small. Because the least squares estimators are unbiased, such that $\mathrm{E}(r_i)$ is zero, a smaller variance of r_i means that the residual r_i is likely to be small or that the fitted value $\widehat{Y}_i$ is likely to be closer to the observed Y_i; that is, the regression line/plane is pulled towards Y_i. Such data points with small residual r_i will not be flagged as outliers by a residual analysis. Instead, they are identified by their large h_{ii} values and are called *high leverage points* exert undue influence on the regression analysis.

According to Corollary 2.6.4,

$$\sum_{i=1}^{n} h_{ii} = k+1 \quad \text{and} \quad 0 \leqslant h_{ii} \leqslant 1.$$

Therefore, two times the average leverage value, namely $\frac{2(k+1)}{n}$, is sometimes used as a benchmark in statistics literature and data points with $h_{ii} > \frac{2(k+1)}{n}$ are classified as high leverage influential points.

10.5 Cook's distance

Another very popular measure for identifying influential observations in regression analysis is *Cook's distance*, which examines the change of $\widehat{\boldsymbol{\beta}}$ that results from deleting an observation from the analysis. Let $\widehat{\boldsymbol{\beta}}_{(i)}$ be the least squares estimate of $\boldsymbol{\beta}$ derived by deleting the i-th observation from the data set in the analysis, then Cook's distance for the i-th observation is given by

$$D_i = \frac{(\widehat{\boldsymbol{\beta}}_{(i)} - \widehat{\boldsymbol{\beta}})^T \mathbf{X}^T \mathbf{X} (\widehat{\boldsymbol{\beta}}_{(i)} - \widehat{\boldsymbol{\beta}})}{(k+1)\widehat{\sigma}^2}. \tag{10.19}$$

Because $\mathrm{Cov}(\widehat{\boldsymbol{\beta}}) = \sigma^2(\mathbf{X}^T\mathbf{X})^{-1}$ (see (10.9)), Cook's distance is a measure of the distance between $\widehat{\boldsymbol{\beta}}_{(i)}$ and $\widehat{\boldsymbol{\beta}}$, which is defined by a notion similar to that used in (8.37). Let $\widehat{\mathbf{Y}}_{(i)} = \mathbf{X}\widehat{\boldsymbol{\beta}}_{(i)}$, then (10.19) reduces to

$$D_i = \frac{(\widehat{\mathbf{Y}}_{(i)} - \widehat{\mathbf{Y}})^T (\widehat{\mathbf{Y}}_{(i)} - \widehat{\mathbf{Y}})}{(k+1)\widehat{\sigma}^2}, \tag{10.20}$$

and Cook's distance is also a measure of the distance between $\widehat{\mathbf{Y}}$ and $\widehat{\mathbf{Y}}_{(i)}$. It is obvious that if D_i is large, then the i-th observation is an influential point because whether or not it is included in the analysis will lead to results that are quite different. We also show (see (10.41)) that D_i is a function of the residual and leverage of the i-th data point. To determine how large, some suggest comparing D_i with an F *distribution* using degrees of freedom $k+1$ and $n-(k+1)$, whereas others simply use a *natural gap* approach and regard data points with D_i values that are substantially larger than those of other data points to be influential.

In the context of a weighted least squares analysis, let $\boldsymbol{\omega} = (\omega_1, \omega_2, \ldots, \omega_n)^T$ be the weights assigned to the n data points. The estimate $\widehat{\boldsymbol{\beta}}_{(i)}$, which is obtained by deleting the i-th observation is a weighted least squares estimate of the form (10.15) and can be obtained by minimizing

$$d_{\boldsymbol{\omega}}(\boldsymbol{\beta}) = \sum_{j=1}^{n} \omega_j (Y_j - (\beta_0 + \beta_1 X_{j1} + \cdots + \beta_k X_{jk}))^2$$

in (10.14) with $\omega_i = 0$ and $\omega_j = 1$ for all $j = 1, ..., n, j \neq i$. In effect, using this more general perspective and considering $\widehat{\boldsymbol{\beta}}_{(i)}$ as a special case of $\widehat{\boldsymbol{\beta}}_{\boldsymbol{\omega}}$ in (10.15) Allow a general form of Cook's distance in (10.19) or (10.20); that

$$D_{\boldsymbol{\omega}} = \frac{(\widehat{\boldsymbol{\beta}}_{\boldsymbol{\omega}} - \widehat{\boldsymbol{\beta}})^T \mathbf{X}^T \mathbf{X} (\widehat{\boldsymbol{\beta}}_{\boldsymbol{\omega}} - \widehat{\boldsymbol{\beta}})}{(k+1)\widehat{\sigma}^2} = \frac{(\widehat{\mathbf{Y}}_{\boldsymbol{\omega}} - \widehat{\mathbf{Y}})^T (\widehat{\mathbf{Y}}_{\boldsymbol{\omega}} - \widehat{\mathbf{Y}})}{(k+1)\widehat{\sigma}^2}, \tag{10.21}$$

where $\widehat{\mathbf{Y}}_{\boldsymbol{\omega}} = \mathbf{X}\widehat{\boldsymbol{\beta}}_{\boldsymbol{\omega}}$. From $\widehat{\mathbf{Y}} = \mathbf{HY}$ (see (10.6)) and the expression of $\widehat{\boldsymbol{\beta}}_{\boldsymbol{\omega}}$ in (10.15), we have

$$\widehat{\mathbf{Y}}_{\boldsymbol{\omega}} = \mathbf{X}\widehat{\boldsymbol{\beta}}_{\boldsymbol{\omega}} = \mathbf{X}(\mathbf{X}^T \boldsymbol{\Omega} \mathbf{X})^{-1} \mathbf{X}^T \boldsymbol{\Omega} \mathbf{Y} = \mathbf{H}_{\boldsymbol{\omega}} \mathbf{Y}, \tag{10.22}$$

where

$$\mathbf{H}_{\boldsymbol{\omega}} = \mathbf{X}(\mathbf{X}^T \boldsymbol{\Omega} \mathbf{X})^{-1} \mathbf{X}^T \boldsymbol{\Omega}. \tag{10.23}$$

As a result, the Cook's distance in (10.21) can also be written as

$$D_{\boldsymbol{\omega}} = \frac{|\mathbf{HY} - \mathbf{H}_{\boldsymbol{\omega}} \mathbf{Y}|^2}{(k+1)\widehat{\sigma}^2}. \tag{10.24}$$

10.6 Deletion influence, joint influence and masking effect

The approach of constructing diagnostic measures by examining the influence of deleting a data point from the analysis is a popular one in influence analysis. Although this kind of intuitively convincing deletion influence measurement is very popular in statistical analyses, it is also well known that measures based on deletion are vulnerable to *masking effects* that may arise in the presence of multiple unusual observations.

The data in Fig. 10.2 show that Observations 1 and 2 are highly influential points in a regression analysis because the regression line has been pulled toward them, as described in Section 10.4. However, changes in the regression line or the parameter estimates due to the deletion of just one observation, say

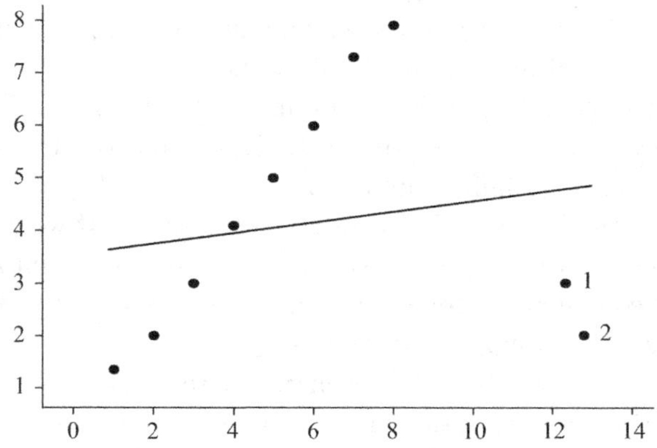

Fig. 10.2 Masking effect in linear regression

Observation 1, are small due to the presence of another extreme observation, Observation 2. The influence of Observation 1, when assessed by measures such as Cook's distance, is masked by that of Observation 2.

To assess the *joint influence* of two or more data points, diagnostic measures derived from deleting a group of cases can be constructed. For example, let $\boldsymbol{\omega} = (\omega_1, \omega_2, \ldots, \omega_n)^T$ with $\omega_i = \omega_j = 0$ and $\omega_k = 1$ for all $k = 1, ..., n, k \neq i, k \neq j$ be the weights assigned to the n data points in a weighted least squares analysis. The estimate $\widehat{\boldsymbol{\beta}}_{(ij)}$, which neglects both the i-th and j-th observations, can be obtained by minimizing $d_{\boldsymbol{\omega}}$ in (10.14) and is a weighted least squares estimate of the form (10.15). The Cook's distance for the i-th and j-th data points, similar to (10.19) and (10.20), can be defined as

$$D_{ij} = \frac{(\widehat{\boldsymbol{\beta}}_{(ij)} - \widehat{\boldsymbol{\beta}})^T \mathbf{X}^T \mathbf{X} (\widehat{\boldsymbol{\beta}}_{(ij)} - \widehat{\boldsymbol{\beta}})}{(k+1)\widehat{\sigma}^2} = \frac{(\widehat{\mathbf{Y}}_{(ij)} - \widehat{\mathbf{Y}})^T (\widehat{\mathbf{Y}}_{(ij)} - \widehat{\mathbf{Y}})}{(k+1)\widehat{\sigma}^2}, \tag{10.25}$$

where $\widehat{\mathbf{Y}}_{(ij)} = \mathbf{X}\widehat{\boldsymbol{\beta}}_{(ij)}$. If D_{ij} is large, then one can conclude that the i-th and j-th data points have a simultaneously large joint influence. We will show (see (C.2)) that D_{ij} can be expressed as a function of the residuals of the i-th and j-th data points and the elements in the hat matrix.

One can also define the Cook's distance D_{ijk} similarly to that in (10.25) to detect the joint influence of the i-th, j-th and k-th cases. In general, the Cook's distances D_i, D_{ij}, D_{ijk}, ... can all be expressed in the form of (10.21); that is,

$$D_{\boldsymbol{\omega}} = \frac{(\widehat{\boldsymbol{\beta}}_{\boldsymbol{\omega}} - \widehat{\boldsymbol{\beta}})^T \mathbf{X}^T \mathbf{X} (\widehat{\boldsymbol{\beta}}_{\boldsymbol{\omega}} - \widehat{\boldsymbol{\beta}})}{(k+1)\widehat{\sigma}^2} = \frac{(\widehat{\mathbf{Y}}_{\boldsymbol{\omega}} - \widehat{\mathbf{Y}})^T (\widehat{\mathbf{Y}}_{\boldsymbol{\omega}} - \widehat{\mathbf{Y}})}{(k+1)\widehat{\sigma}^2}, \tag{10.26}$$

where the weights $\boldsymbol{\omega} = (\omega_1, \omega_2, \ldots, \omega_n)^T$ assigned to the n data points are either 0 or 1, $\widehat{\boldsymbol{\beta}}_{\boldsymbol{\omega}}$ is given by (10.15) and $\widehat{\mathbf{Y}}_{\boldsymbol{\omega}} = \mathbf{X}\widehat{\boldsymbol{\beta}}_{\boldsymbol{\omega}}$.

Although the concept of a group deletion measure is appealing, they are not very popular in practical analysis. If we use the Cook's distance D_i of deleting one data point in (10.19) as a diagnostic measure, then for each data set with size n we must compute n Cook's distances. If we want to detect whether a pair of observations has joint influence by using the measure D_{ij} in (10.25), then we must compute $n(n-1)/2$ D_{ij}'s for a data set of size n. If we use the deletion approach to find whether a group of cases, say m cases, has joint influence, then the number of combinations for which we must compute the measure increases drastically when the group size m increases.

In this deletion approach, the estimates of the regression parameters obtained by deleting one or a group of cases are weighted least squares estimates with a set of appropriately chosen weights $\boldsymbol{\omega} = (\omega_1, \omega_2, \ldots, \omega_n)^T$, which are either 0 or 1. Generalizing this idea of perturbing the weights of observations for weights that are not 0 or 1 allows other measures based on local influence to be constructed. We discuss this further in Part III.

10.7 Derivation of Cook's distances

In this section, we derive the Cook's distance for deleting one data point. The generalization to Cook's distance for deleting two data points is given in Appendix C. We develop the deletion algebra within the context of perturbation to provide a conceptual and empirical comparison between the deletion and perturbation methods introduced in Part III.

10.7.1 Weighted least squares and Cook's distance

We adopt the matrix notation from Section 10.2 for the discussion of weighted least squares (in Section 9.6, leading to (10.14) and (10.15)). For each $\boldsymbol{\omega} = (\omega_1, \ldots, \omega_n)$ with each $\omega_j \geqslant 0$, define

$$d_{\boldsymbol{\omega}}(\boldsymbol{\beta}) = \sum_{i=1}^{n} \omega_i (Y_i - (\beta_0 + \beta_1 X_{i1} + \cdots + \beta_k X_{ik}))^2 \tag{10.27}$$

$$= (\mathbf{Y} - \mathbf{X}\boldsymbol{\beta})^T \boldsymbol{\Omega} (\mathbf{Y} - \mathbf{X}\boldsymbol{\beta}), \tag{10.28}$$

where $\boldsymbol{\Omega} = \text{Diag}(\omega_1, \ldots, \omega_n)$. Let $\mathbf{W} = \text{Diag}(\omega_1^{1/2}, \ldots, \omega_n^{1/2})$, then $\boldsymbol{\Omega} = \mathbf{W}^2$ and

$$d_{\boldsymbol{\omega}}(\boldsymbol{\beta}) = (\mathbf{WY} - \mathbf{WX}\boldsymbol{\beta})^T (\mathbf{WY} - \mathbf{WX}\boldsymbol{\beta}) = |\mathbf{WY} - \mathbf{WX}\boldsymbol{\beta}|^2. \tag{10.29}$$

For a given $\boldsymbol{\omega}$, this quantity is minimized with respect to $\boldsymbol{\beta}$ if we take projection on the column vectors of the matrix $\mathbf{WX}$. Therefore, the projection matrix is

$$\mathbf{WX}(\mathbf{X}^T\mathbf{W}^T\mathbf{WX})^{-1}\mathbf{X}^T\mathbf{W} = \mathbf{WX}(\mathbf{X}^T\boldsymbol{\Omega}\mathbf{X})^{-1}\mathbf{X}^T\mathbf{W}.$$

The minimizer is obtained at

$$\widehat{\boldsymbol{\beta}}_{\boldsymbol{\omega}} = (\mathbf{X}^T\mathbf{W}^T\mathbf{WX})^{-1}\mathbf{X}^T\mathbf{W}^T\mathbf{WY} = (\mathbf{X}^T\boldsymbol{\Omega}\mathbf{X})^{-1}\mathbf{X}^T\boldsymbol{\Omega}\mathbf{Y}. \tag{10.30}$$

The residual is

$$\mathbf{r}_{\boldsymbol{\omega}} = \mathbf{WY} - \mathbf{WX}(\mathbf{X}^T\boldsymbol{\Omega}\mathbf{X})^{-1}\mathbf{X}^T\boldsymbol{\Omega}\mathbf{Y} = \mathbf{W}(\mathbf{I} - \mathbf{X}(\mathbf{X}^T\boldsymbol{\Omega}\mathbf{X})^{-1}\mathbf{X}^T\boldsymbol{\Omega})\mathbf{Y}. \tag{10.31}$$

As in (10.23), let

$$\mathbf{H}_{\boldsymbol{\omega}} = \mathbf{X}(\mathbf{X}^T\boldsymbol{\Omega}\mathbf{X})^{-1}\mathbf{X}^T\boldsymbol{\Omega}, \tag{10.32}$$

and

$$\widehat{\mathbf{Y}}_{\boldsymbol{\omega}} = \mathbf{X}\widehat{\boldsymbol{\beta}}_{\boldsymbol{\omega}} = \mathbf{H}_{\boldsymbol{\omega}}\mathbf{Y}, \tag{10.33}$$

then similar to (10.11),

$$\begin{aligned} d_{\boldsymbol{\omega}}(\widehat{\boldsymbol{\beta}}_{\boldsymbol{\omega}}) &= (\mathbf{Y} - \mathbf{X}\widehat{\boldsymbol{\beta}}_{\boldsymbol{\omega}})^T\boldsymbol{\Omega}(\mathbf{Y} - \mathbf{X}\widehat{\boldsymbol{\beta}}_{\boldsymbol{\omega}}) \\ &= (\mathbf{Y} - \widehat{\mathbf{Y}}_{\boldsymbol{\omega}})^T\boldsymbol{\Omega}(\mathbf{Y} - \widehat{\mathbf{Y}}_{\boldsymbol{\omega}}) \\ &= (\mathbf{Y} - \mathbf{X}(\mathbf{X}^T\boldsymbol{\Omega}\mathbf{X})^{-1}\mathbf{X}^T\boldsymbol{\Omega}\mathbf{Y})^T\boldsymbol{\Omega}(\mathbf{Y} - \mathbf{X}(\mathbf{X}^T\boldsymbol{\Omega}\mathbf{X})^{-1}\mathbf{X}^T\boldsymbol{\Omega}\mathbf{Y}) \\ &= (\mathbf{Y} - \mathbf{H}_{\boldsymbol{\omega}}\mathbf{Y})^T\boldsymbol{\Omega}(\mathbf{Y} - \mathbf{H}_{\boldsymbol{\omega}}\mathbf{Y}). \end{aligned}$$

As a generalization of the discussion in previous sections along with (10.21) and (10.24), for any $\boldsymbol{\omega} = (\omega_1, \ldots, \omega_n)$, the Cook's distance for $\boldsymbol{\omega}$ is defined as

$$D_{\boldsymbol{\omega}} = \frac{1}{(k+1)\widehat{\sigma}^2}|\mathbf{HY} - \mathbf{H}_{\boldsymbol{\omega}}\mathbf{Y}|^2 = \frac{1}{(k+1)\widehat{\sigma}^2}|\widehat{\mathbf{Y}} - \widehat{\mathbf{Y}}_{\boldsymbol{\omega}}|^2, \tag{10.34}$$

where $\widehat{\mathbf{Y}} = \mathbf{HY}$ (see (10.6)) and $\widehat{\mathbf{Y}}_{\boldsymbol{\omega}} = \mathbf{H}_{\boldsymbol{\omega}}\mathbf{Y}$ (see (10.33)).

Lemma 10.7.1 *For any* $\boldsymbol{\omega} = (\omega_1, \ldots, \omega_n)$,

$$D_{\boldsymbol{\omega}} = \frac{1}{(k+1)\widehat{\sigma}^2}(|\mathbf{Y} - \mathbf{H}_{\boldsymbol{\omega}}\mathbf{Y}|^2 - |\mathbf{Y} - \mathbf{HY}|^2). \tag{10.35}$$

Proof: According to do definitions,

$$\mathbf{HH}_{\boldsymbol{\omega}} = \mathbf{X}(\mathbf{X}^T\mathbf{X})^{-1}\mathbf{X}^T\mathbf{X}(\mathbf{X}^T\boldsymbol{\Omega}\mathbf{X})^{-1}\mathbf{X}^T\boldsymbol{\Omega} = \mathbf{X}(\mathbf{X}^T\boldsymbol{\Omega}\mathbf{X})^{-1}\mathbf{X}^T\boldsymbol{\Omega} = \mathbf{H}_{\boldsymbol{\omega}}.$$

Therefore, $\mathbf{HH}_{\boldsymbol{\omega}} = \mathbf{H}_{\boldsymbol{\omega}}$ and $\mathbf{H}_{\boldsymbol{\omega}}^T\mathbf{H} = \mathbf{H}_{\boldsymbol{\omega}}^T\mathbf{H}^T = \mathbf{H}_{\boldsymbol{\omega}}^T$. It follows that

$$\begin{aligned} &(\mathbf{H} - \mathbf{H}_{\boldsymbol{\omega}})^T(\mathbf{H} - \mathbf{H}_{\boldsymbol{\omega}}) \\ &= \mathbf{H}^2 - \mathbf{H}_{\boldsymbol{\omega}}^T\mathbf{H} - \mathbf{HH}_{\boldsymbol{\omega}} + \mathbf{H}_{\boldsymbol{\omega}}^T\mathbf{H}_{\boldsymbol{\omega}} \\ &= \mathbf{H} - \mathbf{H}_{\boldsymbol{\omega}}^T - \mathbf{H}_{\boldsymbol{\omega}} + \mathbf{H}_{\boldsymbol{\omega}}^T\mathbf{H}_{\boldsymbol{\omega}}. \end{aligned} \tag{10.36}$$

On the other hand,

$$\begin{aligned}
&(\mathbf{I}-\mathbf{H}_\omega)^T(\mathbf{I}-\mathbf{H}_\omega)-(\mathbf{I}-\mathbf{H})^T(\mathbf{I}-\mathbf{H})\\
&=(\mathbf{I}-\mathbf{H}_\omega^T-\mathbf{H}_\omega+\mathbf{H}_\omega^T\mathbf{H}_\omega)-(\mathbf{I}-\mathbf{H})\\
&=\mathbf{H}-\mathbf{H}_\omega^T-\mathbf{H}_\omega+\mathbf{H}_\omega^T\mathbf{H}_\omega. \qquad (10.37)
\end{aligned}$$

Because (10.36) is the same as (10.37), (10.34) is the same as (10.35) and the proof is completed. ■

This lemma is useful for relating the *likelihood displacement* of the case-weight perturbation in a liner regression to the Cook's distance, which is addressed in Section 11.3.

10.7.2 Cook's distance – deleting one data point

If we delete one case (e.g. the i-th case) from a linear regression analysis, it is equivalent to considering the weights to be

$$\boldsymbol{\omega}=(1,\ldots,1,0,1,\ldots,1)$$

with 0 in the i-th slot. In this case, $\boldsymbol{\Omega}=\mathbf{I}-\mathbf{e}_i\mathbf{e}_i^T$. Recall that the transpose of the i-th row of the matrix $\mathbf{X}$ is denoted by $\mathbf{X}_i$, i.e.,

$$\mathbf{X}_i=(1,X_{i1},\ldots,X_{ik})^T. \qquad (10.38)$$

Then

$$\mathbf{X}^T\boldsymbol{\Omega}\mathbf{X}=\mathbf{X}^T\mathbf{X}-\mathbf{X}^T\mathbf{e}_i\mathbf{e}_i^T\mathbf{X}=\mathbf{X}^T\mathbf{X}-\mathbf{X}_i\mathbf{X}_i^T.$$

As noted in [?], when $\mathbf{A}$ is a $p\times p$-matrix and $\mathbf{U}$ and $\mathbf{V}$ are $p\times m$-matrices, then

$$(\mathbf{A}-\mathbf{U}\mathbf{V}^T)^{-1}=\mathbf{A}^{-1}+\mathbf{A}^{-1}\mathbf{U}(\mathbf{I}-\mathbf{V}^T\mathbf{A}^{-1}\mathbf{U})^{-1}\mathbf{V}^T\mathbf{A}^{-1}. \qquad (10.39)$$

Therefore, applying the above formula to $\mathbf{A}=\mathbf{X}^T\mathbf{X}$ and $\mathbf{U}=\mathbf{V}=\mathbf{X}_i$,

$$\begin{aligned}
&(\mathbf{X}^T\boldsymbol{\Omega}\mathbf{X})^{-1}\\
&=(\mathbf{X}^T\mathbf{X})^{-1}+(\mathbf{X}^T\mathbf{X})^{-1}\mathbf{X}_i(\mathbf{I}-\mathbf{X}_i^T(\mathbf{X}^T\mathbf{X})^{-1}\mathbf{X}_i)^{-1}\mathbf{X}_i^T(\mathbf{X}^T\mathbf{X})^{-1}\\
&=(\mathbf{X}^T\mathbf{X})^{-1}+\frac{1}{1-h_{ii}}(\mathbf{X}^T\mathbf{X})^{-1}\mathbf{X}_i\mathbf{X}_i^T(\mathbf{X}^T\mathbf{X})^{-1},
\end{aligned}$$

where $h_{ii} = \mathbf{X}_i^T(\mathbf{X}^T\mathbf{X})^{-1}\mathbf{X}_i$ is the leverage of the i-th data point. As a result,

$$\begin{aligned}
&\widehat{\boldsymbol{\beta}}_{(i)} = \widehat{\boldsymbol{\beta}}_{\omega} = (\mathbf{X}^T\boldsymbol{\Omega}\mathbf{X})^{-1}\mathbf{X}^T\boldsymbol{\Omega}\mathbf{Y} \\
&= (\mathbf{X}^T\mathbf{X})^{-1}\mathbf{X}^T\boldsymbol{\Omega}\mathbf{Y} + \frac{1}{1-h_{ii}}(\mathbf{X}^T\mathbf{X})^{-1}\mathbf{X}_i\mathbf{X}_i^T(\mathbf{X}^T\mathbf{X})^{-1}\mathbf{X}^T\boldsymbol{\Omega}\mathbf{Y} \\
&= (\mathbf{X}^T\mathbf{X})^{-1}\mathbf{X}^T\mathbf{Y} - (\mathbf{X}^T\mathbf{X})^{-1}\mathbf{X}_iY_i \\
&\quad + \frac{1}{1-h_{ii}}(\mathbf{X}^T\mathbf{X})^{-1}\mathbf{X}_i\mathbf{X}_i^T(\mathbf{X}^T\mathbf{X})^{-1}\mathbf{X}^T\mathbf{Y} \\
&\quad - \frac{1}{1-h_{ii}}(\mathbf{X}^T\mathbf{X})^{-1}\mathbf{X}_i\mathbf{X}_i^T(\mathbf{X}^T\mathbf{X})^{-1}\mathbf{X}_iY_i.
\end{aligned}$$

Therefore,

$$\begin{aligned}
&\mathbf{HY} - \mathbf{H}_\omega\mathbf{Y} \\
&= \mathbf{X}(\widehat{\boldsymbol{\beta}} - \widehat{\boldsymbol{\beta}}_{(i)}) \\
&= \mathbf{X}(\mathbf{X}^T\mathbf{X})^{-1}\mathbf{X}_iY_i - \frac{1}{1-h_{ii}}\mathbf{X}(\mathbf{X}^T\mathbf{X})^{-1}\mathbf{X}_i\mathbf{X}_i^T(\mathbf{X}^T\mathbf{X})^{-1}\mathbf{X}^T\mathbf{Y} \\
&\quad + \frac{h_{ii}}{1-h_{ii}}\mathbf{X}(\mathbf{X}^T\mathbf{X})^{-1}\mathbf{X}_iY_i \\
&= \frac{1}{1-h_{ii}}\mathbf{X}(\mathbf{X}^T\mathbf{X})^{-1}(\mathbf{X}_iY_i - \mathbf{X}_i\mathbf{X}_i^T(\mathbf{X}^T\mathbf{X})^{-1}\mathbf{X}^T\mathbf{Y}) \\
&= \frac{r_i}{1-h_{ii}}\mathbf{X}(\mathbf{X}^T\mathbf{X})^{-1}\mathbf{X}_i \\
&= \frac{r_i}{1-h_{ii}}\mathbf{He}_i,
\end{aligned}$$

where

$$r_i = Y_i - \mathbf{X}_i^T\widehat{\boldsymbol{\beta}} \tag{10.40}$$

is the residual (see (10.5)) of the i-th data point. It follows that

$$\begin{aligned}
&|\mathbf{HY} - \mathbf{H}_\omega\mathbf{Y}|^2 \\
&= \frac{r_i^2}{(1-h_{ii})^2}\mathbf{X}_i^T(\mathbf{X}^T\mathbf{X})^{-1}\mathbf{X}^T\mathbf{X}(\mathbf{X}^T\mathbf{X})^{-1}\mathbf{X}_i \\
&= \frac{r_i^2 h_{ii}}{(1-h_{ii})^2}.
\end{aligned}$$

According to (10.34), another expression of the Cooks's distance for deleting the i-th case in (10.19) and (10.20) is

$$D_i = \frac{1}{(k+1)\widehat{\sigma}^2}\frac{r_i^2 h_{ii}}{(1-h_{ii})^2}. \tag{10.41}$$

With a similar method, we can derive the Cook's distance of deleting two data points. The computation is presented in Appendix C.

Part III

Local Influence Analysis

Chapter 11

Basic Concepts

11.1 Introduction

Many of the geometric concepts covered in Part I have wide applications in statistics, such as establishing theories in relation to regression, as shown in Part II. In Part III, we see how these concepts can be applied in local influence analyses.

Most statistical analyses require the input of a hypothesized statistical model together with a set of data, and the hypothesized model is only an approximation based on the researchers' best knowledge, which means that it has limitations. Therefore, influence analysis (not confined to "local" influence analysis), which is an assessment of the influence due to perturbation on various aspects of the model inputs, is regarded as an important step in a thorough statistical analysis. Influence analysis results are very informative because they help identify implicit problems in the statistical analysis and judge whether or not the statistical findings and conclusions are reliable.

There are different paradigms for influence analysis in statistics, and local influence analysis is a commonly used approach. The motivation and foundation of the theory of the local influence analysis can be regarded as geometrical. Based on geometric reasonings, many diagnostic measures for statistical analysis can be developed, which lead to a variety of applications in statistics.

One typical example in influence analysis is assessing the influence of an individual data point on the estimation of a statistical model. For example, the Cook's distance (10.19) introduced in Section 10.5 is frequently used in regression and other statistical models to identify data points that may exert undue influence on model estimation. Cook's distance is a so-called *deletion measure* that represents the differences in the least squares estimates of regression coefficients before and after the deletion of a specific data point from the analysis. It is used to judge the contribution of a data point to the determination of the least squares estimates. As suggested in Sections 10.5 to 10.7 in the context of a weighted least squares analysis, the Cook's distance is obtained by perturbing the case weights of the data points, because deleting a data point means displacing its case weight from 1 to 0. In Part III of this book, our focus shifts to *local influence analysis* in which one considers *local perturbation* or *infinitesimal perturbation*. Regarding Cook's distance, local influence analysis deals with the small perturbation of the case weights from the original value 1 as opposed to deleting the case by modifying its weight from 1 to 0.

In the realm of influence analysis, Cook's seminal paper [?] establishes a well-accepted, unified approach for assessing a statistical model from the local perturbation perspective. The approach studies the infinitesimal changes in the log-likelihood of the unperturbed model. A well-known quantifier for extracting information from a perturbation is the normal curvature of a graph, a topic of discussion in Section 4.1. We further explore the use of normal curvature in influence analysis in the next chapter. The local influence approach has been applied in a variety of statistical analyses (e.g., [?, ?, ?, ?, ?, ?, ?, ?, ?, ?]), particularly in assessing the influence of data points in model estimation. In the context of local influence analysis, the core is in terms of the eigenvalues and eigenvectors of the Hessian matrix of the *log-likelihood displacement function*. Over the past twenty years, the utilization of the normal curvature and relevant concepts in local influence analysis have been subjected to further justifications [?, ?, ?] and modifications [?, ?]. For example, Poon and Poon [?] introduced the conformal normal curvature that is analogous to the normal curvature and has the advantage of taking values from 0 to 1, in addition to establishing the concepts such as q-influential and aggregation contributions. These modified works enable the development of objective benchmarks to judge the magnitudes of various measures and provide justification for using or not using more than one eigenvector and the diagonal elements of the Hessian matrix in influence analysis. These generalized concepts have been applied in various studies (e.g., [?, ?, ?, ?, ?, ?, ?, ?, ?, ?, ?, ?, ?]). We will further elaborate, consolidate and extend such concepts in Part III.

Our discussion begins with the perturbation of the likelihood function, the likelihood displacement and the influence graph. The relevant discussion will be illustrated by using a likelihood function given in (11.2). However, the applications are not confined to likelihood function and can be used in many other functions. For example, the concepts of conformal normal curvature, $q-$influential and aggregation contributions [?] have been applied to a conditional expectation of the complete-data, log-likelihood function in the EM algorithm [?].

In Chapter 12, we develop local influence measures for measuring local individual influences and local joint influences. These measures constitute a systematic framework similar to that of the system for the deletion-type influence measures Cook's distances $\{D_i, D_{ij}, D_{ijk}, ...\}$ (see (10.20) and (10.25)). This framework allows for the further investigation of the relations among various measures, the details for which are given in Chapter 13. In Chapter 13, we also explore the importance of eigenvalues beyord the maximum one to establish the concepts of influential eigenvectors and aggregation contribution. In Chapter 14, we consider conformal modifications, including the development of the conformal normal curvature, which has the beneficial property of assuming values between 0 and 1.

11.2 Perturbation

The starting point of a local influence analysis is a sensible perturbation scheme. Perturbation can be represented by a set of parameters collected in $\boldsymbol{\omega} = (\omega_1, \ldots, \omega_n)^T$. The idea of the local influence approach is to introduce appropriate infinitesimal perturbation $\boldsymbol{\omega}$ from the original values $\boldsymbol{\omega}_0$ that represents the original model inputs to assess the impact of some key features of statistical analysis. Statistical considerations inform what perturbation and how it should be introduced to the model. For example, in the linear regression model

$$\mathbf{Y} = \mathbf{X}\boldsymbol{\beta} + \boldsymbol{\epsilon}, \tag{11.1}$$

where $\mathbf{X}$ is an $n \times p$-matrix, $\boldsymbol{\epsilon}$ is an $n \times 1$ normal random vector with $E(\boldsymbol{\epsilon}) = 0$ and $\text{Var}(\boldsymbol{\epsilon}) = \sigma^2\mathbf{I}$, the estimate of $\boldsymbol{\beta}$ is obtained by maximizing the log-likelihood function, as suggested in Sections 9.3 and 10.2. It is a function of $\boldsymbol{\beta}$ and σ^2. Let $\boldsymbol{\theta}$ collect all unknown parameters, including the error variance σ^2 and the regression coefficients in $\boldsymbol{\beta}$. The concerned function is proportional to

$$L(\boldsymbol{\theta}) = -\frac{1}{2\sigma^2}\sum_{i=1}^{n}(Y_i - \mathbf{X}_i^T\boldsymbol{\beta})^2 = -\frac{1}{2\sigma^2}(\mathbf{Y} - \mathbf{X}\boldsymbol{\beta})^T(\mathbf{Y} - \mathbf{X}\boldsymbol{\beta}), \tag{11.2}$$

where Y_i is the i-th entry of $\mathbf{Y}$ and $\mathbf{X}_i$ is the transpose of the i-th row of the matrix $\mathbf{X}$ (see (10.38)). The model in (11.1) or the log-likelihood in (11.2) represents the null model on which the statistical analysis is based.

Let $\boldsymbol{\Omega} = \text{Diag}(\omega_1, \ldots, \omega_n)$ be an $n \times n$-matrix with entries $\omega_1, \ldots, \omega_n$ on the diagonal, then the perturbed log-likelihood for case-weight perturbation is

$$\begin{aligned} L(\boldsymbol{\theta}|\boldsymbol{\omega}) &= -\frac{1}{2\sigma^2}(\mathbf{Y} - \mathbf{X}\boldsymbol{\beta})^T\boldsymbol{\Omega}(\mathbf{Y} - \mathbf{X}\boldsymbol{\beta}) \\ &= -\frac{1}{2\sigma^2}(\mathbf{Y} - \mathbf{X}\boldsymbol{\beta})^T\text{Diag}(\omega_1, \ldots, \omega_n)(\mathbf{Y} - \mathbf{X}\boldsymbol{\beta}) \\ &= -\frac{1}{2\sigma^2}\sum_{i=1}^{n}\omega_i(Y_i - \mathbf{X}_i^T\boldsymbol{\beta})^2. \end{aligned} \tag{11.3}$$

In this example, the perturbation parameters are contained in the diagonal matrix $\boldsymbol{\Omega}$. We recover the unperturbed model when $\boldsymbol{\omega} = \boldsymbol{\omega}_0 = (1, \ldots, 1)^T$, i.e., $\text{Diag}(\omega_1, \ldots, \omega_n) = \mathbf{I}$. The local influence approach involves infinitesimal perturbations of the weights. However, if $\omega_1 = 0$ and $\omega_j = 1$ for all $j \geqslant 2$, then the first data point is deleted from analysis. One may consider the case deletion of the Cook's distance, as presented in Section 10.5, to be a form of perturbation of the weights of the cases from 1 to 0, which establishes the connection between the deletion influence and the local influence approaches.

We will use this case-weight perturbation in regression (11.3) for illustration throughout our discussion of local influence. Case-weight perturbation is the most popular perturbation scheme in local influence analyses, because identifying influential data points that exert undue influence on a statistical analysis is an important issue. For example, recall that the sample mean $\overline{X} = (\sum X_i)/n$ is used in many statistical analyses to provide information on the location of the data and is commonly used to estimate the population mean μ (see Sections 6.3 and 6.3.1). It is obvious that a data point with extreme value that is far away from the others will highly influence the value of the sample mean, which produces an estimate that is not reliable. It is thus of crucial importance to identify such data points with extreme values. While such identification may be straightforward when there is only one variable, it becomes more difficult in problems with more variables. An approach based on case-weight perturbation can be employed to develop diagnostic measures [?].

There are many ways other than case-weights to introduce a perturbation scheme and different perturbation schemes will lead to diagnostic measures that are sensitive to different features of the model inputs. Statistical considerations inform how to introduce a sensible perturbation scheme and a variety of useful perturbations are available in the literature (see, e.g., [?, ?, ?, ?, ?, ?, ?, ?, ?, ?, ?]).

11.3 Likelihood displacement and influence graph

Because the method of maximum likelihood (see Section 6.5) is a typical method for finding the estimates of model parameters, a perturbation of the model inputs in many cases can be represented by a perturbation of the *log-likelihood function*. In effect, the case-weight perturbation in (11.3) is a perturbation of the log-likelihood function of the regression model. In general, given the log-likelihood $L(\boldsymbol{\theta})$ for a postulated statistical model with $\boldsymbol{\theta} = (\theta_1, \ldots, \theta_p)^T$ as a vector of p unknown parameters, we let $L(\boldsymbol{\theta}|\boldsymbol{\omega})$ be the log-likelihood of a perturbed model with perturbation parameter $\boldsymbol{\omega} = (\omega_1, \ldots, \omega_n)^T$ defined in an open subset $\mathcal{U}$ in $\mathbb{R}^n$. Let $\boldsymbol{\omega}_0$ be a point of *null perturbation* in $\mathcal{U}$, such that

$L(\boldsymbol{\theta}|\boldsymbol{\omega}_0) = L(\boldsymbol{\theta})$. For any given $\boldsymbol{\omega}$, let $\widehat{\boldsymbol{\theta}}_{\boldsymbol{\omega}}$ be the maximizer of $L(\boldsymbol{\theta}|\boldsymbol{\omega})$. It is found as a solution of the equation

$$\frac{\partial L(\boldsymbol{\theta}|\boldsymbol{\omega})}{\partial \omega_j} = 0$$

for all $1 \leqslant j \leqslant n$. The *log-likelihood displacement* or *likelihood displacement* is a function of the perturbation parameters, given by

$$f(\boldsymbol{\omega}) = 2\left(L(\widehat{\boldsymbol{\theta}}_{\boldsymbol{\omega}_0}|\boldsymbol{\omega}_0) - L(\widehat{\boldsymbol{\theta}}_{\boldsymbol{\omega}}|\boldsymbol{\omega}_0)\right). \tag{11.4}$$

By construction, $f(\boldsymbol{\omega}_0) = 0$ because the unperturbed log-likelihood achieves its maximum when $\boldsymbol{\omega} = \boldsymbol{\omega}_0$. Because of the local nature of our consideration, we assume that the domain $\mathcal{U}$ is so small that $f(\boldsymbol{\omega}) \geqslant 0$ for all $\boldsymbol{\omega} \in \mathcal{U}$. In other words, f attains a minimum at the unperturbed point $\boldsymbol{\omega}_0$.

Likelihood displacement is widely used in the classical influence analysis literature [?] because it provides a general framework that can unify the concepts of different influence or diagnostic measures and can be applied to a variety of statistical models. In the context of the local influence approach and linear regression with case-weight perturbation, we can see from the following that the likelihood displacement is equivalent to a "locally-perturbed" version of the Cook's distance. According to (11.2), the likelihood displacement $f(\boldsymbol{\omega})$ in (11.4) becomes

$$\begin{aligned}
&2\left(L(\widehat{\boldsymbol{\theta}}_{\boldsymbol{\omega}_0}|\boldsymbol{\omega}_0) - L(\widehat{\boldsymbol{\theta}}_{\boldsymbol{\omega}}|\boldsymbol{\omega}_0)\right) \\
&= 2\left(\left(-\frac{1}{2\sigma^2}(\mathbf{Y} - \mathbf{X}\widehat{\boldsymbol{\beta}})^T(\mathbf{Y} - \mathbf{X}\widehat{\boldsymbol{\beta}})\right) - \left(-\frac{1}{2\sigma^2}(\mathbf{Y} - \mathbf{X}\widehat{\boldsymbol{\beta}}_{\boldsymbol{\omega}})^T(\mathbf{Y} - \mathbf{X}\widehat{\boldsymbol{\beta}}_{\boldsymbol{\omega}})\right)\right)
\end{aligned}$$

where $\widehat{\boldsymbol{\beta}}$ is the maximum likelihood estimate of the regression parameters under the unperturbed model as given by (10.4), and $\widehat{\boldsymbol{\beta}}_{\boldsymbol{\omega}}$ maximizes the log-likelihood (11.3) of the perturbed model as given by (10.30). Because $\widehat{\mathbf{Y}} = \mathbf{X}\widehat{\boldsymbol{\beta}}$, $\widehat{\mathbf{Y}}_{\boldsymbol{\omega}} = \mathbf{X}\widehat{\boldsymbol{\beta}}_{\boldsymbol{\omega}}$, the above quantity becomes

$$-\frac{1}{\sigma^2}\left(|\mathbf{Y} - \widehat{\mathbf{Y}}|^2 - |\mathbf{Y} - \widehat{\mathbf{Y}}_{\boldsymbol{\omega}}|^2\right).$$

According to (10.34) and Lemma 10.7.1, it is equal to

$$\frac{1}{\sigma^2}\left(|\widehat{\mathbf{Y}} - \widehat{\mathbf{Y}}_{\boldsymbol{\omega}}|^2\right).$$

It becomes clear that the likelihood displacement in (11.4) is related to the Cook's distance.

As noted in Section 3.1, the graph G of the function $f(\boldsymbol{\omega})$ is the image of a map from $\mathcal{U}$ into $\mathbb{R}^{n+1}$:

$$\phi(\boldsymbol{\omega}) = \begin{pmatrix} \boldsymbol{\omega} \\ f(\boldsymbol{\omega}) \end{pmatrix}. \tag{11.5}$$

This is called the *influence graph* of the likelihood displacement function $f(\boldsymbol{\omega})$. By studying the behavior of the influence graph near $\boldsymbol{\omega}_0$, we apply the local influence approach to obtain an understanding of the influence of a chosen perturbation scheme, which allows us to assess the vulnerability of the findings of a statistical analysis to minor perturbations of the inputs for that analysis. We introduce several tools for assessing such influence and the relationships among the tools in subsequent chapters.

Chapter 12

Measuring Local Influence

Recall our discussion in Section 10.5 on Cook's distance D_i (10.20), which assesses the influence of deleting a single case from the analysis, and the generalizations to D_{ij} in (10.25) found in Section 10.6 for assessing joint influence based on the deletion of two data points as well as $D_{\boldsymbol{\omega}}$ with an appropriately chosen $\boldsymbol{\omega}$ in (10.26) for the multiple deletion of a group of data points. In each of these assessments, we are interested in the effectiveness of the measures used to assess the influence and understanding of the fundamental relations among different assessment measures.

From $\{D_i, D_{ij}, D_{ijk}, \ldots\}$ to measures based on the multiple deletion of a group of cases, Cook's distances measure deletion influence. This entire series of measures can be expressed as $D_{\boldsymbol{\omega}}$, where the entries of $\boldsymbol{\omega}$ are either zero or one (see (10.26) and Section 10.7). The concept behind this series of measures and their uses are intuitively convincing despite the computational ineffectiveness that arises when handling a large number of combinations [?]. Comparing multiple-deletion and single-deletion Cook's distances make it easy to examine and understand the limitations of the single-deletion Cook's distance [?]. Similarly, when assessing local influence due to infinitesimal perturbations, one may assess the influence of an individual perturbation parameter or a pair or group of perturbation parameters. We establish local influence measures that can be incorporated into a systematic framework similar to that of the system for deletion influence $\{D_i, D_{ij}, D_{ijk}, \ldots\}$, which makes local influence measures as convincing and appealing as deletion measures.

In this chapter, we establish various tools for measuring local influence using geometric techniques. The development of measures is based on Cook's work [?] and a subsequent analysis [?]. In the next chapter, we analyze their relations and efficiencies.

12.1 Individual influence

A perturbation model is determined by the log-likelihood displacement function $f(\boldsymbol{\omega})$ which is defined on a small neighborhood $\mathcal{U}$ of the unperturbed point $\boldsymbol{\omega}_0$. The influence graph G of the function $f(\boldsymbol{\omega})$ over $\mathcal{U}$ can be treated as the image of the map ϕ (see Section 3.1 and expression (11.5)). As Section 3.2 shows, for any given direction $\boldsymbol{\ell}$ in the tangent space of the graph over the point $\boldsymbol{\omega}_0$, we get a normal section $G(\boldsymbol{\ell})$ on the graph. If we consider the perturbation of $\boldsymbol{\omega}_0$ only in the parameter ω_j and hold all other parameters unchanged, then the log-likelihood displacement function becomes an exclusive function of ω_j. Its graph is precisely the normal section $G(\boldsymbol{e}_j)$. Consider the quantity

$$C_j = C_{\mathbf{e}_j}, \tag{12.1}$$

the normal curvature of the cross section $G(\boldsymbol{e}_j)$ of the graph in the plane that is spanned by the normal vector $\mathbf{n}$ and basic vector $\mathbf{e}_j$ in $\mathbb{R}^{n+1}$. Because the curve $G(\boldsymbol{e}_j)$ can also be considered the graph of the likelihood displacement function obtained by holding all of the perturbation parameters, with the exception of ω_j as constant and unperturbed, C_j measures the sensitivity of a single parameter of perturbation. By comparing C_i with C_j, we detect the relative sensitivity of the two individual parameters. This perspective frames C_j as a measure of the influence, which is due to the perturbation of a single parameter. A priori, it does not carry any information on how the j-th perturbation parameter is related to the other parameters. Therefore, we consider C_j the *local individual influence* due to the parameter ω_j.

We can extend the consideration of the individual influence of each parameter ω_j to include the consideration of the individual influence of each linear combination $\boldsymbol{\ell}$, which is determined using the normal curvature $C_{\boldsymbol{\ell}}$ of the graph $G(\boldsymbol{\ell})$ in the direction $\boldsymbol{\ell}$.

According to Proposition 4.1.1, the normal curvature of graph G at point $\phi(\boldsymbol{\omega})$ in direction $\boldsymbol{\ell}$ is given in terms of the first and second fundamental forms and is equal to

$$C_{\boldsymbol{\ell}} = \frac{\mathrm{II}(\boldsymbol{\ell},\boldsymbol{\ell})}{\mathbb{I}(\boldsymbol{\ell},\boldsymbol{\ell})}. \tag{12.2}$$

As noted in (3.16) and (4.6),

$$\mathbb{I}(\boldsymbol{\ell},\boldsymbol{\ell}) = \boldsymbol{\ell}^T(\mathbf{I} + {\nabla_f}^T\nabla_f)\boldsymbol{\ell}, \qquad \mathrm{II}(\boldsymbol{\ell},\boldsymbol{\ell}) = \frac{\boldsymbol{\ell}^T\ddot{\mathbf{F}}\boldsymbol{\ell}}{\sqrt{1+\nabla_f^T\nabla_f}},$$

where ∇_f is the gradient of the function f and $\ddot{\mathbf{F}}$ is the Hessian matrix of the function f at the point $\boldsymbol{\omega}$.

Because the unperturbed point $\boldsymbol{\omega}_0$ is a minimum of the likelihood function f, the gradient of the function f at $\boldsymbol{\omega}_0$ is equal to zero and the Hessian matrix becomes semi-positive definite at this point. Equivalently, all eigenvalues are non-negative. In particular, at $\boldsymbol{\omega}_0$

$$C_{\boldsymbol{\ell}} = \frac{\boldsymbol{\ell}^T \ddot{\mathbf{F}} \boldsymbol{\ell}}{\boldsymbol{\ell}^T \boldsymbol{\ell}} = \frac{\langle \ddot{\mathbf{F}} \boldsymbol{\ell}, \boldsymbol{\ell} \rangle}{\langle \boldsymbol{\ell}, \boldsymbol{\ell} \rangle}. \tag{12.3}$$

From this expression, it is also clear that for any non-zero constant t,

$$C_{t\boldsymbol{\ell}} = C_{\boldsymbol{\ell}}. \tag{12.4}$$

Therefore, to compute normal curvatures, we consider $\boldsymbol{\ell}$ as subjected to the condition $|\boldsymbol{\ell}| = 1$ such that,

$$C_{\boldsymbol{\ell}} = \boldsymbol{\ell}^T \ddot{\mathbf{F}} \boldsymbol{\ell} = \langle \ddot{\mathbf{F}} \boldsymbol{\ell}, \boldsymbol{\ell} \rangle. \tag{12.5}$$

In particular,

$$C_j = C_{\mathbf{e}_j} = \mathbf{e}_j^T \ddot{\mathbf{F}} \mathbf{e}_j = \frac{\partial^2 f}{\partial \omega_j^2}. \tag{12.6}$$

When the function f in question is the log-likelihood displacement function of a perturbation model in applied statistics, as described in Section 11.3, the normal curvature C_{ℓ} is sometimes addressed as the*Cook's directional curvature* or simply *Cook's curvature* [?].

12.2 Derivation of normal curvature

As noted in the last section, the computation of the normal curvature function is in terms of the first and second derivative of the log-likelihood displacement function $f(\boldsymbol{\omega})$, pareicularly at the unperturbed point $\boldsymbol{\omega}_0$. In this section, we exhibit some calculations that facilitate applications and future discussions.

Following the notation in Section 11.3, letting $\widehat{\theta}_{\boldsymbol{\omega},a}$ be the a-th element of $\widehat{\boldsymbol{\theta}}_{\boldsymbol{\omega}}$ and applying the Chain Rule and implicit differentiation of the function $f(\boldsymbol{\omega})$ in (11.4), we have

$$\frac{\partial f}{\partial \omega_j} = -2 \sum_{a=1}^{p} \frac{\partial L(\boldsymbol{\theta}|\boldsymbol{\omega}_0)}{\partial \theta_a} \Big|_{\boldsymbol{\theta}=\widehat{\boldsymbol{\theta}}_{\boldsymbol{\omega}}} \frac{\partial \widehat{\theta}_{\boldsymbol{\omega},a}}{\partial \omega_j}. \tag{12.7}$$

For all $\boldsymbol{\omega}$, $\widehat{\boldsymbol{\theta}}_{\boldsymbol{\omega}}$ is a solution to the equation for all a,

$$\frac{\partial L(\boldsymbol{\theta}|\boldsymbol{\omega})}{\partial \theta_a} = 0.$$

Specifically,

$$\frac{\partial L(\boldsymbol{\theta}|\boldsymbol{\omega})}{\partial \theta_a}\Big|_{\boldsymbol{\theta}=\widehat{\boldsymbol{\theta}}_{\boldsymbol{\omega}}} = 0. \tag{12.8}$$

In particular, the above identity holds for $\boldsymbol{\omega}_0$. Therefore, one sees from (12.7) that $\nabla_f(\boldsymbol{\omega}_0) = \mathbf{0}$.

Given that $\widehat{\boldsymbol{\theta}}_{\boldsymbol{\omega}}$ is a function of $\boldsymbol{\omega}$, we take a derivative of (12.7) to obtain

$$\begin{aligned}\frac{\partial^2 f}{\partial \omega_i \partial \omega_j} &= -2\sum_{a,b=1}^{p} \frac{\partial^2 L(\boldsymbol{\theta}|\boldsymbol{\omega}_0)}{\partial \theta_a \partial \theta_b}\Big|_{\boldsymbol{\theta}=\widehat{\boldsymbol{\theta}}_{\boldsymbol{\omega}}} \frac{\partial \widehat{\theta}_{\boldsymbol{\omega},b}}{\partial \omega_i}\frac{\partial \widehat{\theta}_{\boldsymbol{\omega},a}}{\partial \omega_j} \\ &\quad -2\sum_{a=1}^{p} \frac{\partial L(\boldsymbol{\theta}|\boldsymbol{\omega}_0)}{\partial \theta_a}\Big|_{\boldsymbol{\theta}=\widehat{\boldsymbol{\theta}}_{\boldsymbol{\omega}}} \frac{\partial^2 \widehat{\theta}_{\boldsymbol{\omega},a}}{\partial \omega_i \partial \omega_j}.\end{aligned} \tag{12.9}$$

Due to (12.8), the evaluation of the second term at $\boldsymbol{\omega}_0$ is equal to zero.

Define a $p \times p$ matrix $\ddot{\mathbf{L}}$ and a $p \times n$ matrix $\mathbf{J}$ by

$$\ddot{\mathbf{L}}_{ab} = \frac{\partial^2 L(\boldsymbol{\theta}|\boldsymbol{\omega}_0)}{\partial \theta_a \partial \theta_b}\Big|_{\boldsymbol{\theta}=\widehat{\boldsymbol{\theta}}}, \qquad \mathbf{J}_{bj} = \frac{\partial \widehat{\theta}_{\boldsymbol{\omega},b}}{\partial \omega_j}\Big|_{\boldsymbol{\omega}=\boldsymbol{\omega}_0}. \tag{12.10}$$

Then the evaluation of (12.9) yields

$$\ddot{\mathbf{F}} = -2\mathbf{J}^T\ddot{\mathbf{L}}\mathbf{J}. \tag{12.11}$$

On the other hand, we take the derivative of the identity in (12.8) with respect to ω_j to get

$$\frac{\partial^2 L(\boldsymbol{\theta}|\boldsymbol{\omega})}{\partial \theta_a \partial \theta_b}\Big|_{\boldsymbol{\theta}=\widehat{\boldsymbol{\theta}}_{\boldsymbol{\omega}}} \frac{\partial \widehat{\theta}_{\boldsymbol{\omega},b}}{\partial \omega_j} + \frac{\partial^2 L(\boldsymbol{\theta}|\boldsymbol{\omega})}{\partial \theta_a \partial \omega_j}\Big|_{\boldsymbol{\theta}=\widehat{\boldsymbol{\theta}}_{\boldsymbol{\omega}}} = 0. \tag{12.12}$$

Define a $p \times n$ matrix $\boldsymbol{\Delta}$ by

$$\boldsymbol{\Delta}_{aj} = \frac{\partial^2 L(\boldsymbol{\theta}|\boldsymbol{\omega}_0)}{\partial \theta_a \partial \omega_j}\Big|_{\boldsymbol{\theta}=\widehat{\boldsymbol{\theta}}}. \tag{12.13}$$

Evaluate (12.12) at $\boldsymbol{\omega} = \boldsymbol{\omega}_0$. Since $\widehat{\boldsymbol{\theta}} = \widehat{\boldsymbol{\theta}}_{\boldsymbol{\omega}_0}$, we obtain $\ddot{\mathbf{L}}\mathbf{J} + \boldsymbol{\Delta} = 0$. Equivalently,

$$\mathbf{J} = -\ddot{\mathbf{L}}^{-1}\boldsymbol{\Delta}.$$

Substitute it in (12.11), we have

$$\ddot{\mathbf{F}} = -2\mathbf{J}^T\ddot{\mathbf{L}}\mathbf{J} = -2\boldsymbol{\Delta}^T\ddot{\mathbf{L}}^{-1}\ddot{\mathbf{L}}\ddot{\mathbf{L}}^{-1}\boldsymbol{\Delta} = -2\boldsymbol{\Delta}^T\ddot{\mathbf{L}}^{-1}\boldsymbol{\Delta}. \tag{12.14}$$

According to (12.14) and (12.3), when $\boldsymbol{\ell}^T\boldsymbol{\ell} = 1$

$$C_{\boldsymbol{\ell}} = \boldsymbol{\ell}^T\ddot{\mathbf{F}}\boldsymbol{\ell} = -2\boldsymbol{\ell}^T\boldsymbol{\Delta}^T\ddot{\mathbf{L}}^{-1}\boldsymbol{\Delta}\boldsymbol{\ell}. \tag{12.15}$$

12.3 Case-weight perturbation — an example

Consider the case-weight perturbation (11.3) of a simple linear regression model. Assuming σ^2 is known, then the unknown parameters are the regression coefficients $\boldsymbol{\beta}$; that is, $\boldsymbol{\theta} = \boldsymbol{\beta}$. In this case and from (11.3), the perturbed likelihood function is determined by the function

$$L(\boldsymbol{\theta}|\boldsymbol{\omega}) = -\frac{1}{2\sigma^2}(\mathbf{Y} - \mathbf{X}\boldsymbol{\theta})^T\boldsymbol{\Omega}(\mathbf{Y} - \mathbf{X}\boldsymbol{\theta}). \tag{12.16}$$

Therefore,

$$\begin{aligned}\frac{\partial L}{\partial \theta_a} &= \frac{1}{2\sigma^2}\Big((\mathbf{X}\boldsymbol{e}_a)^T\boldsymbol{\Omega}(\mathbf{Y} - \mathbf{X}\boldsymbol{\theta}) + (\mathbf{Y} - \mathbf{X}\boldsymbol{\theta})^T\boldsymbol{\Omega}\mathbf{X}\boldsymbol{e}_a\Big)\\ &= \frac{1}{2\sigma^2}\Big(\boldsymbol{e}_a^T\mathbf{X}^T\boldsymbol{\Omega}(\mathbf{Y} - \mathbf{X}\boldsymbol{\theta}) + (\mathbf{Y} - \mathbf{X}\boldsymbol{\theta})^T\boldsymbol{\Omega}\mathbf{X}\boldsymbol{e}_a\Big), \qquad (12.17)\\ \frac{\partial^2 L}{\partial \theta_a \partial \theta_b} &= -\frac{1}{\sigma^2}\boldsymbol{e}_a^T\mathbf{X}^T\boldsymbol{\Omega}\mathbf{X}\boldsymbol{e}_b. \qquad (12.18)\end{aligned}$$

This means that

$$\ddot{\mathbf{L}} = -\frac{1}{\sigma^2}\mathbf{X}^T\mathbf{X}, \quad \text{and} \quad \ddot{\mathbf{L}}^{-1} = -\sigma^2(\mathbf{X}^T\mathbf{X})^{-1}. \tag{12.19}$$

In addition, if we the derivative of (12.17) with respect to ω_j, then we get

$$\begin{aligned}\frac{\partial^2 L}{\partial \theta_a \partial \omega_j} &= \frac{1}{2\sigma^2}\Big(\boldsymbol{e}_a^T\mathbf{X}^T\boldsymbol{e}_j\boldsymbol{e}_j^T(\mathbf{Y} - \mathbf{X}\boldsymbol{\theta}) + (\mathbf{Y} - \mathbf{X}\boldsymbol{\theta})^T\boldsymbol{e}_j\boldsymbol{e}_j^T\mathbf{X}\boldsymbol{e}_a\Big)\\ &= \frac{1}{\sigma^2}(Y_j - \mathbf{X}_j^T\boldsymbol{\theta})\mathbf{X}_{j,a}\\ &= \frac{1}{\sigma^2}r_j\mathbf{X}_{j,a},\end{aligned}$$

where r_j is the residual (see (10.5) to (10.8)). In other words, when $\boldsymbol{\theta} = \widehat{\boldsymbol{\theta}}$,

$$\boldsymbol{\Delta} = \frac{1}{\sigma^2}\mathbf{X}^T\mathbf{R} \tag{12.20}$$

where $\mathbf{R} = \text{Diag}(r_1, \ldots, r_n)$ is the diagonal matrix whose entries are the coordinates of the residual vector $\mathbf{r}$. Substitute (12.19) and (12.20) in (12.14), and we find that

$$\ddot{\mathbf{F}} = \frac{2}{\sigma^2}\mathbf{R}^T\mathbf{X}(\mathbf{X}^T\mathbf{X})^{-1}\mathbf{X}^T\mathbf{R} = \frac{2}{\sigma^2}\mathbf{RHR}, \tag{12.21}$$

where $\mathbf{H} = \mathbf{X}(\mathbf{X}^T\mathbf{X})^{-1}\mathbf{X}^T$ is the hat matrix .

According to (12.15)

$$C_{\boldsymbol{\ell}} = \frac{2}{\sigma^2}\boldsymbol{\ell}^T\mathbf{RHR}\boldsymbol{\ell}. \tag{12.22}$$

Let the (i,j)-th entry for the Hessian matrix $\ddot{\mathbf{F}}$ of f be denoted by f_{ij} and that for the hat matrix $\mathbf{H}$ as h_{ij}, then equation (12.21) means that

$$f_{ij} = \frac{2}{\sigma^2} r_i h_{ij} r_j. \tag{12.23}$$

According to (12.22), for each basic vector $\boldsymbol{e}_j$, we have

$$C_j = C_{\boldsymbol{e}_j} = \frac{2}{\sigma^2} r_j^2 h_{jj}. \tag{12.24}$$

To illustrate the above ideas, consider the Data Set A in Table 12.1. Using (12.24), one may calculate the normal curvature for each basic vector $\{\boldsymbol{e}_1, \ldots, \boldsymbol{e}_j, \ldots, \boldsymbol{e}_n\}$. In practical analysis, when σ^2 is unknown, one can replace σ^2 with its maximum likelihood estimate $\widehat{\sigma}^2_{ML}$ (see (10.12)). The theoretical justification for doing so can be found in [?] and [?]. We exhibit the result in Figure 12.1 with an index plot, i.e., to plot the value of the normal curvature against the index $j \in \{1, \ldots, n\}$.

Table 12.1 Data set A

Case	1	2	3	4	5	6	7	8	9	10
x	1.0	2.0	3.0	4.0	5.0	6.0	7.0	8.0	9.0	10.0
y	1.5	3.6	3.8	5.2	6.2	6.4	7.6	9.3	10.2	11.0

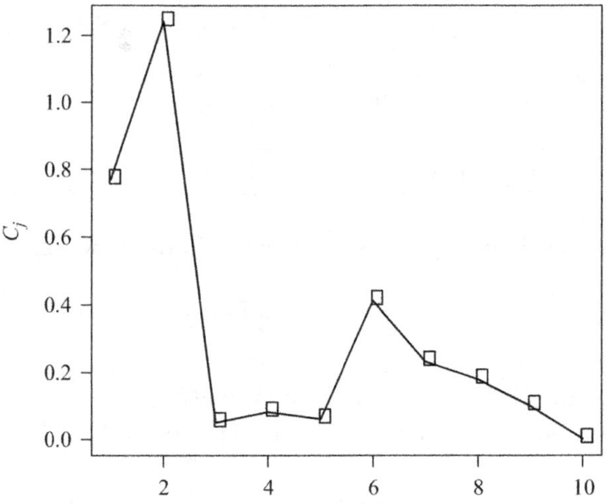

Fig. 12.1 Index plot of Cook's curvature for data set A

As C_1 and C_2 are large compared to the other C_j's, the influence of these two perturbation parameters is more substantial. In other words, the likelihood function changes more substantially if the weights of Case 1 and Case 2 are displaced slightly from 1, Which suggests that Cases 1 and 2 are influential points.

12.4 Roles of sectional curvature

In the last few sections, we develop the concept and measures of local individual influence. In this and the next section, we develop the concept of local joint influence and search of its measure.

In developing the concept of joint influence and an associated measure for a pair of perturbation parameters, we attempt to apply two different ideas. The first is a natural and direct generalization of the idea of normal curvature from a geometric perspective. The second returns to Cook's quest for an infinitesimal combination with maximal sensitivity. Once we identify a way to assess the influence of a pair of perturbation parameters, it is easy to generalize the concept and computational tools to a group of multiple perturbation parameters.

We begin with a pair of perturbation parameters and seek measures on them at an infinitesimal level. From a mathematical perspective it is natural to expand Cook's geometric consideration of one-dimensional to two-dimensional cross sections, as given in Section 3.3.

Recall that given any pair of linearly independent tangent vectors $\boldsymbol{\ell}$ and $\boldsymbol{\ell}'$ of the influence graph at $\boldsymbol{\omega}_0$, the three-dimensional span of $\boldsymbol{\ell}, \boldsymbol{\ell}'$ and the normal vector $\mathbf{n}$ cuts across the graph of the log-likelihood displacement function along a two-dimensional surface. Denote the cross section of the graph of the log-likelihood displacement in this three-dimensional space as $G(\boldsymbol{\ell}, \boldsymbol{\ell}')$.

If the two vectors being studied are $\boldsymbol{e}_i$ and $\boldsymbol{e}_j$ in the standard basis, the cross section is the graph of function f with respect to the perturbation parameters ω_i and ω_j, $G(\boldsymbol{e}_i, \boldsymbol{e}_j)$.

As seen in Section 4.2, sectional curvature is conceived as a generalization of normal curvature. For a general, two-dimensional cross section such as $G(\boldsymbol{\ell}, \boldsymbol{\ell}')$, its sectional curvature $\mathrm{Sec}(\boldsymbol{\ell}, \boldsymbol{\ell}')$ is equal to

$$\mathrm{Sec}(\boldsymbol{\ell}, \boldsymbol{\ell}') = \frac{\langle \boldsymbol{\ell}, \ddot{\mathbf{F}}\boldsymbol{\ell}\rangle\langle \boldsymbol{\ell}', \ddot{\mathbf{F}}\boldsymbol{\ell}'\rangle - \langle \boldsymbol{\ell}, \ddot{\mathbf{F}}\boldsymbol{\ell}'\rangle\langle \boldsymbol{\ell}', \ddot{\mathbf{F}}\boldsymbol{\ell}\rangle}{\langle \boldsymbol{\ell}, \boldsymbol{\ell}\rangle\langle \boldsymbol{\ell}', \boldsymbol{\ell}'\rangle - \langle \boldsymbol{\ell}, \boldsymbol{\ell}'\rangle^2}. \tag{12.25}$$

In particular, based on Proposition 4.2.2, $\mathrm{Sec}(\boldsymbol{e}_i, \boldsymbol{e}_j) = \det \ddot{\mathbf{F}}_{(ij)}$ where

$$\ddot{\mathbf{F}}_{(ij)} = \begin{pmatrix} \dfrac{\partial^2 f}{\partial \omega_i^2} & \dfrac{\partial^2 f}{\partial \omega_i \partial \omega_j} \\ \dfrac{\partial^2 f}{\partial \omega_i \partial \omega_j} & \dfrac{\partial^2 f}{\partial \omega_j^2} \end{pmatrix}. \tag{12.26}$$

This observation means that for the graph of the log-likelihood displacement function, the sectional curvature at the unperturbed point $\boldsymbol{\omega}_0$ with respect to the two parameters ω_i and ω_j is the Hessian of the log-likelihood displacement function with only two variables: ω_i and ω_j.

De spite its standard geometric generalization, sectional curvature is not an effective measure of joint influence. The following example illustrates the problem. Recall formula (12.21) for the case-weight perturbation of the linear regression model:

$$\ddot{\mathbf{F}}_{(ij)} = \frac{2}{\sigma^2}\begin{pmatrix} r_i^2 h_{ii} & r_i h_{ij} r_j \\ r_j h_{ij} r_i & r_j^2 h_{jj} \end{pmatrix}. \tag{12.27}$$

Therefore,

$$\mathrm{Sec}(\boldsymbol{e}_i, \boldsymbol{e}_j) = \det \ddot{\mathbf{F}}_{(ij)} = \frac{2}{\sigma^2}\left(r_i^2 r_j^2 (h_{ii}h_{jj} - h_{ij}^2)\right).$$

In the regression model, it is very common for two cases, say the i-th case and the j-th case, to have replicated data. In particular, when the values of the X variables are the same, the i-th row and the j-th row of the hat matrix are identical, i.e., $h_{ik} = h_{jk}$ for all k. As a result, $\mathrm{Sec}(\boldsymbol{e}_i, \boldsymbol{e}_j) = 0$. As $\mathrm{Sec}(\boldsymbol{\ell}, \boldsymbol{\ell}')$ is a continuous function of $\boldsymbol{\ell}$ and $\boldsymbol{\ell}'$, whenever the values of the X variables for the i-th case and the j-th case are close to each other, $\mathrm{Sec}(\boldsymbol{e}_i, \boldsymbol{e}_j)$ is close to zero, which disregards the potentially influential nature of this pair of data points. This observation demonstrates the ineffectiveness of sectional curvature and the limitations of a straightforward geometric generalization of normal curvature.

To reinforce this point with a concrete example, consider the following data set.

Table 12.2 Data set B

Case	1	2	3	4	5	6	7	8	9	10	11	12
x	1.0	2.0	3.0	4.0	5.0	6.0	7.0	8.0	9.0	10.0	14.0	14.0
y	1.5	3.6	3.8	5.2	6.2	6.4	7.6	9.3	10.2	11.0	21.0	20.5

The data and the regression line plot is given below in Figure 12.2.

Intuitively, Cases 11 and 12 are influential. Indeed, as noted in the index plot in Figure 12.3 for the pair-wise Cook's distance where the index k in the horizontal axis is given by $k = \frac{1}{2}(i-1)(i-2) + j$, they form an influential pair.

However, sectional curvature completely fails to detect the influence for this pair of data points, as shown in the plot of $\mathrm{Sec}_{ij} = \mathrm{Sec}(\mathbf{e}_i, \mathbf{e}_j)$ in Figure 12.4.

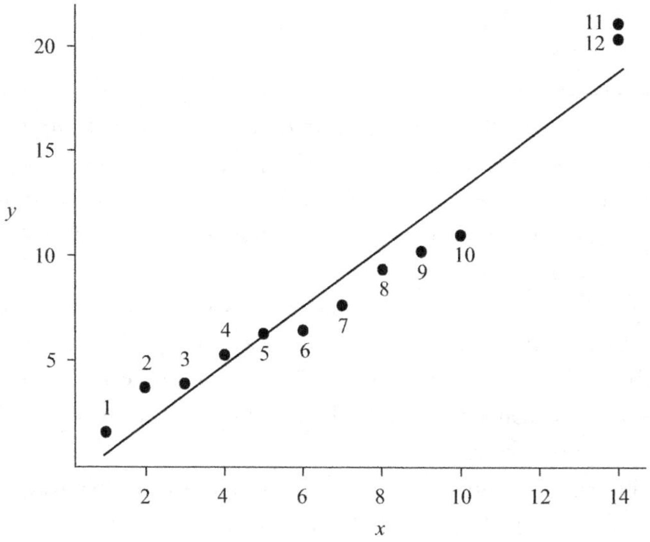

Fig. 12.2 Regression line for data set B

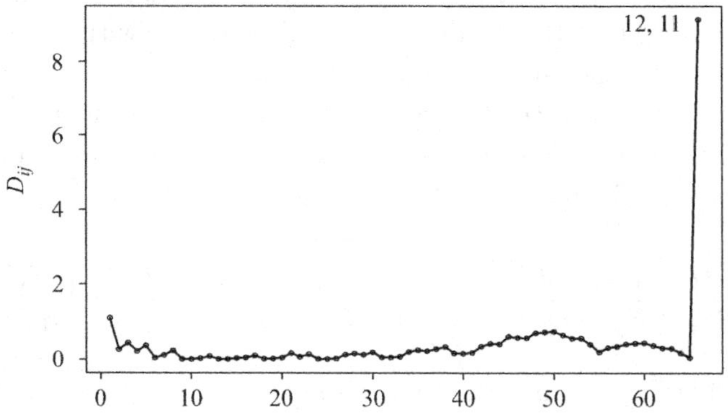

Fig. 12.3 Pairwise Cook's distance D_{ij}, indexed by $k = \frac{1}{2}(i-1)(i-2)+j$

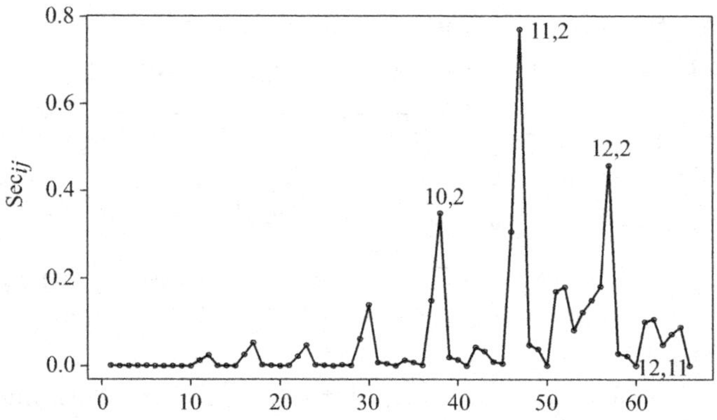

Fig. 12.4 Sectional curvature $\mathrm{Sec}_{ij} = \mathrm{Sec}(\mathbf{e}_i, \mathbf{e}_j)$, indexed by $k = \frac{1}{2}(i-1)(i-2)+j$

12.5 Joint influence

In the last section, we illustrate that sectional curvature is ineffective as a measure of the joint influence of a pair of perturbation parameters. To seek an appropriate measure, we recall that the purpose of Cook's consideration of normal curvature is to identify the local influence of the perturbation parameters. The key idea is to identify the parameters that are subject to maximal sensitivity to perturbation.

When we study the two-dimensional cross section $G(\boldsymbol{e}_i, \boldsymbol{e}_j)$, we are not interested in its overall curvatures such as the information given by sectional curvature. Instead, within this cross section, our objective is to identify the direction among all concerned perturbation parameters with maximum sensitivity or maximum normal curvature.

As noted in the previous section, the two-dimensional cross section $G(\boldsymbol{e}_i, \boldsymbol{e}_j)$ is the graph of the log-likelihood displacement function with respect to the perturbation parameters ω_i and ω_j. The Hessian at the unperturbed point $\boldsymbol{\omega}_0$ is given by the 2×2 matrix $\ddot{\mathbf{F}}_{(ij)}$. Applying Proposition 1.2.3 to the symmetric matrix $\ddot{\mathbf{F}}$, we find that the maximum normal curvature with the linear span of $\boldsymbol{e}_i$ and $\boldsymbol{e}_j$ occurs along an eigenvector of the matrix $\ddot{\mathbf{F}}_{(ij)}$. The maximum value is precisely the larger of the two eigenvalues of the 2×2 matrix $\ddot{\mathbf{F}}_{(ij)}$.

Because the matrix $\ddot{\mathbf{F}}_{(ij)}$ is given by (12.26), it is an elementary exercise to verify that its maximum eigenvalue is given by

$$J_{ij} = \frac{1}{2}\left(\frac{\partial^2 f}{\partial \omega_i^2} + \frac{\partial^2 f}{\partial \omega_j^2}\right) + \sqrt{\frac{1}{4}\left(\frac{\partial^2 f}{\partial \omega_i^2} - \frac{\partial^2 f}{\partial \omega_j^2}\right)^2 + \left(\frac{\partial^2 f}{\partial \omega_i \partial \omega_j}\right)^2}. \qquad (12.28)$$

We consider J_{ij}, given in (12.28), as the *local joint influence* of the i-th and j-th parameters.

As noted in the last section, for the linear regression model with case weight perturbation, the matrix $\ddot{\mathbf{F}}_{(ij)}$ is given in terms of residuals and entries of the hat matrix $\mathbf{H}$ (12.27). Therefore, (12.28) can be reduced to

$$\begin{aligned} J_{ij} &= \frac{1}{\sigma^2}\left(r_i^2 h_{ii} + r_j^2 h_{jj} + \sqrt{(r_i^2 h_{ii} - r_j^2 h_{jj})^2 + 4(r_i h_{ij} r_j)^2}\right) \\ &= \frac{1}{\sigma^2}\left(r_i^2 h_{ii} + r_j^2 h_{jj} + \sqrt{(r_i^2 h_{ii} + r_j^2 h_{jj})^2 - 4 r_i^2 r_j^2 (h_{ii} h_{jj} - h_{ij}^2)}\right). \end{aligned}$$

For the Data Set B, given in Table 12.2, the pairwise local joint influences J_{ij} are given in the index plot of Figure 12.5, where the index k is given by $k = \frac{1}{2}(i-1)(i-2) + j$.

Here we see that the pairs of the 11th and 12th and 10th and 11th data points, respectively, are marked as most influential when measured by J_{ij}.

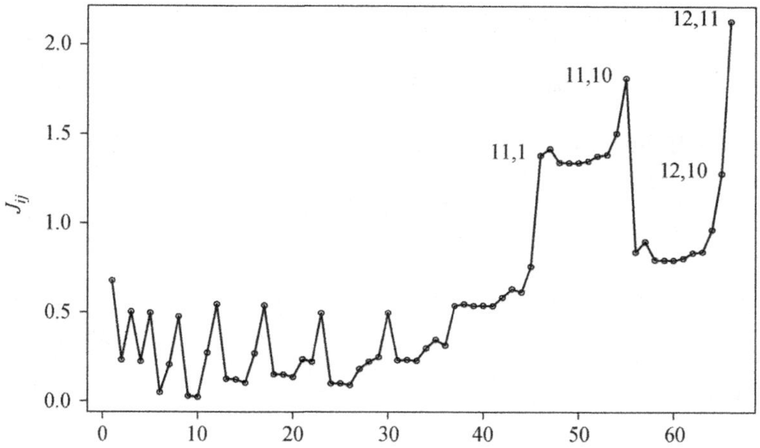

Fig. 12.5 Pairwise local joint influence J_{ij} for data set B

Minor changes in the weights of data points 11 and 12 and 10 and 11 have a substantial influence on the regression parameters estimates. The joint local influence of data points 10 and 11 cannot be identified if one relies solely on pairwise Cook's distance, as seen in Figure 12.3.

If we are interested in the joint influence of a group of more than two parameters, e.g., $\omega_{i_1}, \ldots, \omega_{i_k}$, we consider the cross section of the graph determined by the linear span of the normal vector $\mathbf{n}$ and the basic vectors $\boldsymbol{e}_{i_1}, \ldots, \boldsymbol{e}_{i_k}$. The Hessian of the perturbation of the first k parameters is simply the restriction of the Hessian $\ddot{\mathbf{F}}$ to the first k rows and k columns. The local joint influence of the parameters $\omega_{i_1}, \ldots, \omega_{i_k}$ is the maximum eigenvalue of the $k \times k$-minor of $\ddot{\mathbf{F}}$, given by the i_1-th, ..., i_k-th columns and rows, which is denoted by $J_{i_1,\ldots,i_k}$.

For example, one can begin with the first basic vector $\boldsymbol{e}_1$ and consider its cross section $G(\boldsymbol{e}_1)$. The normal curvature of this cross section is C_1 and measures the local influence of the perturbation parameter ω_1. For the joint influence of the parameters ω_1 and ω_2, we consider the cross section $G(\boldsymbol{e}_1, \boldsymbol{e}_2)$, the joint influence which is J_{12}. This is the maximum eigenvalue of the 2×2-matrix, given by the first two rows and first two columns of the matrix $\ddot{\mathbf{F}}$.

If we consider $\{\boldsymbol{e}_1, \boldsymbol{e}_2, \boldsymbol{e}_3\}$, then the joint influence J_{123} is the maximum eigenvalue of the 3×3 matrix, given by the first three rows and first three columns of the matrix $\ddot{\mathbf{F}}$. This process is generalized to $\{\boldsymbol{e}_1, \ldots, \boldsymbol{e}_k\}$, and it terminates when we involve all of the vectors in the standard basis $\{\boldsymbol{e}_1, \ldots, \boldsymbol{e}_n\}$. In such case, the *overall joint influence* $J_{12\cdots n}$ is precisely the maximum eigenvalue $\lambda_{\max}$ of the full matrix $\ddot{\mathbf{F}}$.

In all of our subsequent discussion, eigenvectors and eigenvalues of $\ddot{\mathbf{F}}$ play a crucial role. Denote and order the eigenvalues of the Hessian of the log-likelihood displacement function by

$$\lambda_{\max} = \lambda_1 \geqslant \lambda_2 \geqslant \cdots \geqslant \lambda_n = \lambda_{\min}. \tag{12.29}$$

The corresponding set of orthonormal eigenvectors is denoted by

$$\mathbf{v}_{\max} = \mathbf{v}_1, \mathbf{v}_2, \cdots, \mathbf{v}_n = \mathbf{v}_{\min}. \tag{12.30}$$

Chapter 13

Relations Among Various Measures

In practical applications, we are often interested in identifying the perturbation parameters that have a large individual influence or a large joint influence exclusively among the basic directions $e_1, \ldots, e_n$. In such a case, one seeks the maximum among $\{C_1, \ldots, C_n\}$ for individual influence or the maximum among all possible J_{ij}, with $1 \leqslant i \neq j \leqslant n$ for pairwise joint influence, and so on.

It is quite practical to find the maximum among $\{C_1, \ldots, C_n\}$, because it is simply the diagonal term of the Hessian matrix $\ddot{\mathbf{F}}$ of the log-likelihood displacement function. To find the maximum amongst all possible $J_{i_1 \cdots i_k}$ for each $k \geqslant 2$, theoretically speaking, one would simply calculate all $J_{i_1 \cdots i_k}$ and then identify their maximums. However, in some applications, the analytical expression of J is difficult to find. Even when the expression can be derived, the large number of possible combinations when k and n are large hinders the feasibility of finding all $J_{i_1 \cdots i_k}$, which limits and hence the applicability of the measures. This is similar to the situation in which group-deletion-type measures cannot gain popularity in statistical analysis. Nonetheless, it is important to know the properties of the various local influence measures and their relations, because doing so facilitates applications. This is similar to the quest for gaining an understanding of the limitations of single deletion measures by examining the relations between single and multiple deletion measures [?]. We begin by finding the range of the local influence measures.

13.1 A bound on influence measures

Recall equation (12.4), in which for any non-zero constant t, $C_{t\boldsymbol{\ell}} = C_{\boldsymbol{\ell}}$. This means that the value of $C_{\boldsymbol{\ell}}$ depends exclusively on the line of direction, but does not depend on the choice of vectors in a particular line. It also means that to find all possible values of $C_{\boldsymbol{\ell}}$, one needs only evaluate $C_{\boldsymbol{\ell}}$ for all directions $\boldsymbol{\ell}$

subjected to the condition $|\boldsymbol{\ell}|^2 = 1$, which reduces the evaluation of $C_{\boldsymbol{\ell}}$ to

$$C_{\boldsymbol{\ell}} = \boldsymbol{\ell}^T \ddot{\mathbf{F}} \boldsymbol{\ell}$$

for all $\boldsymbol{\ell}$ with $|\boldsymbol{\ell}|^2 = 1$. Because $\ddot{\mathbf{F}}$ is a symmetric matrix, according to Proposition 1.2.3, the maximum value of $C_{\boldsymbol{\ell}}$ is equal to the maximum eigenvalue $\lambda_{\max}$ of the Hessian matrix $\ddot{\mathbf{F}}$ at the unperturbed point $\boldsymbol{\omega}_0$ and is attained at the eigenvector $\mathbf{v}_{\max}$, which corresponds to $\lambda_{\max}$. Thus, $C_j \leqslant \lambda_{\max}$ for all j. A similar argument works for finding the minimum. Therefore, all individual local influences are bounded by the maximum and minimum eigenvalues of the Hessian matrix; that is,

$$\lambda_{\min} \leqslant C_1, \ldots, C_n \leqslant \lambda_{\max}. \tag{13.1}$$

For the data given in Table 12.1, the rank of the Hessian matrix is equal to 2 (see Appendix A), and its eigenvalues are, respectively,

$$\lambda_{\max} = \lambda_1 = 2.2890, \quad \lambda_2 = 0.8419, \quad \lambda_3 = \cdots = \lambda_{10} = \lambda_{\min} = 0. \tag{13.2}$$

Then, all individual influence measures take place within the interval $[0, 2.2890]$.

A bound on pairwise local joint influence J_{ij} can be similarly determined. When we consider the local joint influence between two perturbation parameters ω_i and ω_j, we note the two-dimensional cross section $G(\boldsymbol{e}_i, \boldsymbol{e}_j)$. Now, for any two perturbation vectors $\boldsymbol{\ell}$ and $\boldsymbol{\ell}'$, we can consider their associated cross section $G(\boldsymbol{\ell}, \boldsymbol{\ell}')$ of the full graph. Within this cross section we study any linear combination $\boldsymbol{e} = a\boldsymbol{\ell} + b\boldsymbol{\ell}'$ with their associated cross section $G(\boldsymbol{e})$. Because cross section $G(\boldsymbol{e})$ within the linear span of $\boldsymbol{\ell}$ and $\boldsymbol{\ell}'$ is also the cross section of the entire graph, the normal curvature of $G(\boldsymbol{e})$ as a subspace of $G(\boldsymbol{\ell}, \boldsymbol{\ell}')$ is simply the normal curvature of $G(\boldsymbol{e})$ in the full graph, which is equal to $C_{\boldsymbol{e}}$. Therefore, the *local joint influence* of the combinations of $\boldsymbol{\ell}$ and $\boldsymbol{\ell}'$ is the quantity

$$J(\boldsymbol{\ell}, \boldsymbol{\ell}') = \max\{C_{\boldsymbol{e}} : \boldsymbol{e} = a\boldsymbol{\ell} + b\boldsymbol{\ell}'\}.$$

In the above definition of $J(\boldsymbol{\ell}, \boldsymbol{\ell}')$, we take the maximum within the linear span of $\boldsymbol{\ell}$ and $\boldsymbol{\ell}'$. Therefore, the quantity $J(\boldsymbol{\ell}, \boldsymbol{\ell}')$ is independent of the choice of basis within the same linear span. In other words, if the linear span of $\mathbf{v}$ and $\mathbf{v}'$ is equal to the span of $\boldsymbol{\ell}$ and $\boldsymbol{\ell}'$, then $J(\mathbf{v}, \mathbf{v}') = J(\boldsymbol{\ell}, \boldsymbol{\ell}')$.

When we search for the maximum $J(\boldsymbol{\ell}, \boldsymbol{\ell}')$ for all possible $\boldsymbol{\ell}$ and $\boldsymbol{\ell}'$, we search for the maximum $C_{\boldsymbol{e}}$ for all possible vector $\boldsymbol{e}$ with unit length, which is equal to $\lambda_{\max}$ and is achieved in the direction of the first eigenvector $\mathbf{v}_{\max} = \mathbf{v}_1$ of the Hessian matrix $\ddot{\mathbf{F}}$.

Similarly, the minimum J_{ij} cannot be smaller than the minimum of $J(\boldsymbol{\ell}, \boldsymbol{\ell}')$ for all possible $\boldsymbol{\ell}$ and $\boldsymbol{\ell}'$. An argument similar to that presented above rereals that the minimum of $J(\boldsymbol{\ell}, \boldsymbol{\ell}')$ is at least as big as the minimum eigenvalue of the Hessian matrix. Therefore, the joint influence measure J_{ij}, like the individual influence measure C_j, fluctuates between $\lambda_{\min}$ and $\lambda_{\max}$. In particular, we have the following bound for pairwise joint influence

$$\lambda_{\min} \leqslant J_{ij} \leqslant \lambda_{\max}.$$

If one is interested in the maximal triple joint influence amongst any possible combinations, then one seeks the maximum of $J(\boldsymbol{\ell}_1, \boldsymbol{\ell}_2, \boldsymbol{\ell}_3)$ for an arbitrary choice of $\boldsymbol{\ell}_1, \boldsymbol{\ell}_2$ and $\boldsymbol{\ell}_3$. Therefore, the combination $a_1\boldsymbol{\ell}_1 + a_2\boldsymbol{\ell}_2 + a_3\boldsymbol{\ell}_3$ is an arbitrary vector in the tangent space of the graph over $\boldsymbol{\omega}_0$ and, the maximum normal curvature is, again, the maximum eigenvalue of the Hessian matrix, which is attained in the direction of $\mathbf{v}_1$.

This argument can be repeated to search for the maximum of $J(\boldsymbol{\ell}_{i_1}, \ldots, \boldsymbol{\ell}_{i_k})$ for any number k between 1 and n. Thus, when one considers individual influence or joint influence without placing any constraints on the combinations of the perturbation parameters, the maximum is always equal to $\lambda_{\max}$ and is attained in the direction of $\mathbf{v}_{\max}$ of eigenvector with eigenvalue $\lambda_{\max}$.

It follows that for all $i_i, \ldots, i_k$,

$$\lambda_{\min} \leqslant J(\boldsymbol{\ell}_{i_1}, \ldots, \boldsymbol{\ell}_{i_k}) \leqslant \lambda_{\max}, \tag{13.3}$$

such that the overall joint influence of all perturbation parameters $J_{12\cdots n}$ is equal to $\lambda_{\max}$.

13.2 Individual and overall joint influence

To understand the underlying structures and potential relations among the local influence measures with the aim of facilitating the better use of all of the measures, we analyze the relations among C_j and various joint influence measures in this and the next section. This is facilitated by the eigenvectors of the Hessian of the log-likelihood displacement function at $\boldsymbol{\omega}_0$.

We again use the notations given in (12.29) and (12.30) for eigenvalues and eigenvectors and express the eigenvector $\mathbf{v}_i$ as a linear combination of the basic vectors $\boldsymbol{e}_j$; that is, $\mathbf{v}_i = \sum_{j=1}^{n} a_{ji}\boldsymbol{e}_j$. Consider the matrix

$$\mathbf{A} = (\mathbf{v}_1, \mathbf{v}_2, \ldots, \mathbf{v}_n) = \begin{pmatrix} a_{11} & \cdots & a_{1n} \\ \vdots & & \vdots \\ a_{n1} & \cdots & a_{nn} \end{pmatrix}. \tag{13.4}$$

Since we choose $\{\mathbf{v}_i; 1 \leqslant i \leqslant n\}$ as an orthonormal basis, matrix $\mathbf{A}$ is orthogonal. Therefore, $\mathbf{A}^{-1} = \mathbf{A}^T$. According to the construction $\mathbf{A}\boldsymbol{e}_j = \mathbf{v}_j$,

$\mathbf{A}^T\mathbf{v}_j = \mathbf{A}^{-1}\mathbf{v}_j = \boldsymbol{e}_j$, i.e.,

$$\boldsymbol{e}_j = \sum_{i=1}^{n} a_{ji}\mathbf{v}_i. \tag{13.5}$$

Let $\mathbf{\Lambda} = \mathrm{Diag}(\lambda_1, \lambda_2, \ldots, \lambda_n)$ be the diagonal matrix of eigenvalues, then collect all of the identities $\ddot{\mathbf{F}}\mathbf{v}_j = \lambda_j\mathbf{v}_j$ in matrix form and we get the equation $\ddot{\mathbf{F}}\mathbf{A} = \mathbf{A}\mathbf{\Lambda}$. Therefore,

$$\ddot{\mathbf{F}} = \mathbf{A}\mathbf{\Lambda}\mathbf{A}^{-1} = \mathbf{A}\mathbf{\Lambda}\mathbf{A}^T.$$

It follows that

$$\frac{\partial^2 f}{\partial\omega_i\partial\omega_j}(\boldsymbol{\omega}_0) = \sum_{k=1}^{n}(\lambda_k a_{ik} a_{jk}) \tag{13.6}$$

and

$$C_j = \frac{\partial^2 f}{\partial\omega_j^2}(\boldsymbol{\omega}_0) = \sum_{k=1}^{n}(\lambda_k a_{jk}^2). \tag{13.7}$$

Suppose there is a single dominant eigenvalue for the Hessian matrix $\ddot{\mathbf{F}}$. For all $k \geqslant 2$, λ_k/λ_1 is very small and the trace of the matrix $\ddot{\mathbf{F}}$ is approximated by $\lambda_{\max}$. By definition of the trace and normal curvature C_j, we have

$$C_1 + C_2 + \cdots + C_n = \mathrm{Tr}\,\ddot{\mathbf{F}} \cong \lambda_{\max}. \tag{13.8}$$

The overall joint influence is approximated by the sum of the individual influence.

In addition, given (13.7), there is an approximation.

$$\frac{C_j}{\lambda_1} \cong a_{j1}^2. \tag{13.9}$$

Because a_{j1} are the coordinates of the eigenvector $\mathbf{v}_1 = \mathbf{v}_{\max}$, the order of the measures $C_1, \ldots, C_n$ coincides with the order of the magnitude of the coefficients in the eigenvector $\mathbf{v}_{\max}$, which brings forth the role of $\mathbf{v}_{\max}$ in measuring individual local influence.

Theorem 13.2.1 *When the Hessian $\ddot{\mathbf{F}}$ of the log-likelihood displacement function has a single dominant eigenvalue, the order of the individual influence C_j agrees with the order of the magnitude of coordinates of the eigenvector $\mathbf{v}_{\max}$ with a relation approximated by* (13.9).

For example, consider the Data Set C given in Table 13.1.

Table 13.1 Data set C

Case	1	2	3	4	5	6	7	8	9	10	11	12	13	14
x	1.0	2.0	3.0	4.0	5.0	6.0	7.0	8.0	9.0	10.0	14.0	14.0	14.0	14.0
y	1.5	3.6	3.8	5.2	6.2	6.4	7.6	9.3	10.2	11.0	21.0	20.5	9.5	9.0

The first eigenvalue λ_1 is equal to 4.7309, and the second eigenvalue λ_2 is equal to 0.0312. They differ by order of magnitude. Now, we provide an index plot of the normal curvature in Figure 13.1 and an index plot of the coefficients of the eigenvector for the eigenvalue λ_1 in Figure 13.2.

These plots vividly reveal the relation between C_j and the coefficients in $\mathbf{v}_{\max}$.

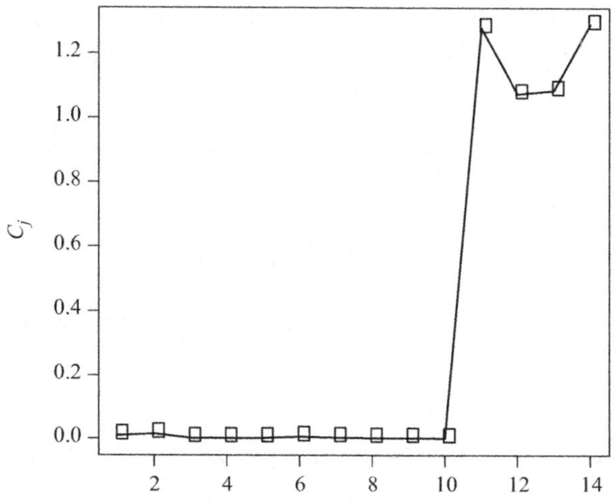

Fig. 13.1 Index plot of C_j for data set C

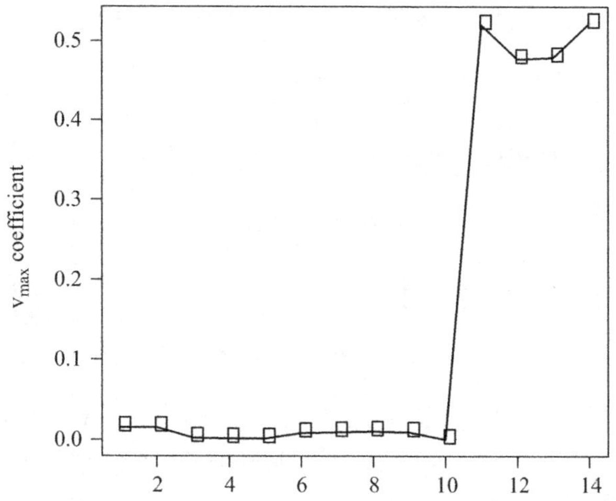

Fig. 13.2 Index plot of coefficient in $\mathbf{v}_{\max}$ for data set C

13.3 Individual and joint influence measures

We continue with our analysis under the assumption that there is a single dominant eigenvalue and analyze the relation between the individual influences C_j and the joint influence measures such as J_{ij}, J_{ijk}.

Under the current assumption, according to (13.6)

$$\frac{\partial^2 f}{\partial\omega_i \partial\omega_j}(\boldsymbol{\omega}_0) = \sum_{k=1}^{n}(\lambda_k a_{ik} a_{jk}) \cong \lambda_{\max} a_{i1} a_{j1}. \tag{13.10}$$

Substitute it in (12.28) and, after some elementary algebra, we find that J_{ij} is approximated by $\lambda_{\max}(a_{i1}^2 + a_{j1}^2)$:

$$J_{ij} \cong \lambda_{\max}(a_{i1}^2 + a_{j1}^2). \tag{13.11}$$

In contrast, according to (13.9) $C_j \cong \lambda_{\max} a_{j1}^2$. We conclude that

$$J_{ij} \cong C_i + C_j. \tag{13.12}$$

This observation can be further generalized. Note that given the notation in (13.4), the coordinates of the vector $\mathbf{v}_{\max}$ are

$$\mathbf{v}_{\max} = (a_{11}, \ldots, a_{n1})^T.$$

Therefore, the matrix form of the approximation (13.10) is

$$\ddot{\mathbf{F}} \cong \lambda_{\max} \mathbf{v}_{\max} \mathbf{v}_{\max}^T \tag{13.13}$$

where we use the convention in Section 1.1.5; that is,

$$\lambda_{\max} \mathbf{v}_{\max} \mathbf{v}_{\max}^T = \lambda_{\max} \begin{pmatrix} a_{11}a_{11} & a_{11}a_{21} & \cdots & a_{11}a_{n1} \\ a_{21}a_{11} & a_{21}a_{21} & \cdots & a_{21}a_{n1} \\ \vdots & \vdots & & \vdots \\ a_{n1}a_{11} & a_{n1}a_{21} & \cdots & a_{n1}a_{n1} \end{pmatrix}. \tag{13.14}$$

As noted in Section 1.1.5, the matrix $\lambda_{\max} \mathbf{v}_{\max} \mathbf{v}_{\max}^T$ is a rank one matrix. Its eigenvalue is obtained by evaluating this matrix on the vector $\mathbf{v}_{\max}$ and is equal to

$$\lambda_{\max}(a_{11}^2 + a_{21}^2 + \cdots + a_{n1}^2) = \lambda_{\max}|\mathbf{v}_{\max}|^2 = \lambda_{\max}.$$

Now, if one wants to compute J_{123}, the joint influence of the first three perturbation parameters, then one must compute the maximum eigenvalue of the upper left 3×3 matrix in $\ddot{\mathbf{F}}$. Under the current assumption, it is approximated by

$$\lambda_{\max} \begin{pmatrix} a_{11}a_{11} & a_{11}a_{21} & a_{11}a_{31} \\ a_{21}a_{11} & a_{21}a_{21} & a_{21}a_{31} \\ a_{31}a_{11} & a_{31}a_{21} & a_{31}a_{31} \end{pmatrix}.$$

The only eigenvalue of this matrix is obtained by evaluating this matrix on $(a_{11}, a_{21}, a_{31})^T$, which is equal to $\lambda_{\max}(a_{11}^2 + a_{21}^2 + a_{31}^2)$. Given (13.9), we have the approximation

$$J_{123} \cong C_1 + C_2 + C_3. \tag{13.15}$$

An identical argument works for J_{1234} and up to $J_{12\ldots n}$. After permuting the indices, one find that this argument works for all combinations of indices. Therefore, we find the generalization of the observation in (13.8).

Theorem 13.3.1 *When the Hessian $\ddot{\mathbf{F}}$ of the log-likelihood displacement function at ω_0 has a single dominant eigenvalue, the joint influence measures $J_{i_1 i_2 \ldots i_k}$ are approximated by the sum of the individual influences C_{i_1}, $C_{i_2}, \ldots,$ C_{i_k}.*

13.4 Competing eigenvalues

Although the majority of our attention has been on the eigenvector $\mathbf{v}_{\max}$, theoretically speaking it is possible for some of the eigenvalues of the Hessian matrix of the log-likelihood function to repeat each other. For example, one might have

$$\lambda_{\max} = \lambda_1 = \lambda_2.$$

In this case, every vector in the linear combination of $\mathbf{v}_1$ and $\mathbf{v}_2$ is an eigenvector with the eigenvalue $\lambda_{\max}$. This leads to the ambiguity of the candidates for $\mathbf{v}_{\max}$ in our analysis in the last two sections because it is not unique.

Of course, in practical problems one is unlikely to encounter a theoretical perfect equality such as $\lambda_1 = \lambda_2$. However, it is possible that the magnitude of λ_1 and λ_2 are comparable. Given the implication of the theoretical extreme, we should manage and involve both $\mathbf{v}_1$ and $\mathbf{v}_2$ in our analysis with care as long as λ_1 and λ_2 are comparable.

For example, take Data Set B from Table 12.2 with case-weight perturbation in regression. An index plot of normal curvature is given in Figure 13.3.

This indicates that data points 11 and 12 are influential, and date Data points 2 and 10 are modestly and comparably influential. For this data set, the first eigenvalue of the log-likelihood displacement function $\lambda_1 = \lambda_{\max}$ is equal

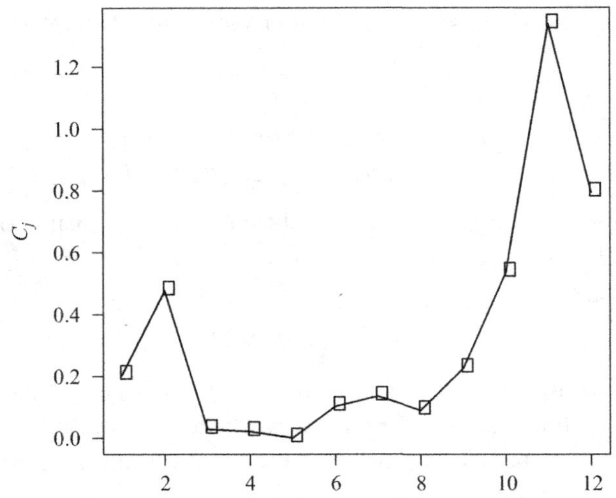

Fig. 13.3 Index plot of C_j for data set B

to 2.9485. An index plot of the coefficients of the first eigenvector $\mathbf{v}_1 = \mathbf{v}_{\max}$ with respect to the standard basis is given in Figure 13.4.

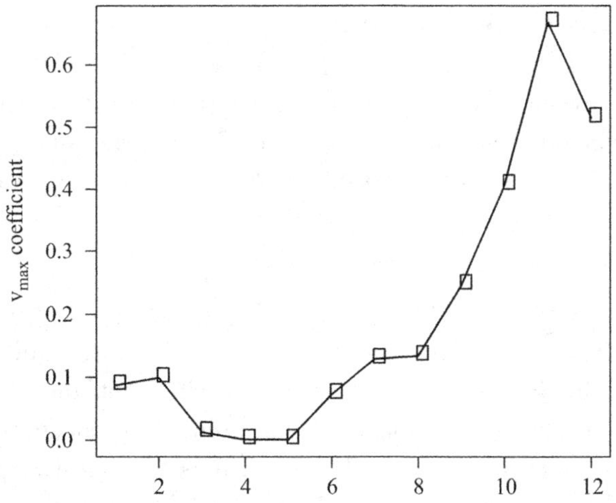

Fig. 13.4 Index plot of coefficient in $\mathbf{v}_{\max}$ for data set B

Although Figure 13.4 confirms that the leading roles of data points 11 and 12 and the potential role of data point 10, data point 2 makes an insignificant contribution to $\mathbf{v}_{\max}$. The disparity between using C_j and coefficients of $\mathbf{v}_{\max}$ resides in the second data point's leading contribution to the second eigenvector, as seen in Figure 13.5. More importantly, the second eigenvalue λ_2 is equal to 1.0004, which is not negligible when compared to the first eigenvalue.

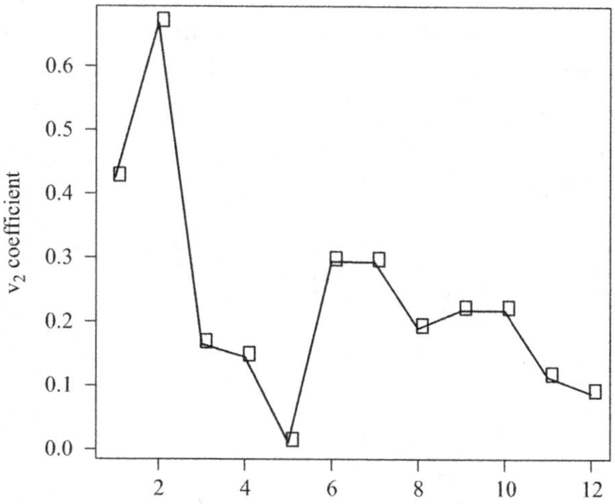

Fig. 13.5 Index plot of coefficient in $\mathbf{v}_2$ for data set B

Such consideration prompts the need to decide when we should investigate the roles of eigenvalues and eigenvectors beyond the maximum.

Recall that the average of the eigenvalues of the Hessian matrix $\ddot{\mathbf{F}}$ of the log-likelihood function is also known as the mean curvature $\overline{\lambda}$ (see Section 4.1). If all eigenvalues are identical, then they are all equal to the average, which is denoted by $\overline{\lambda}$. It is known that the *mean curvature* is equal to

$$\overline{\lambda} = \frac{1}{n} \operatorname{Tr} \ddot{\mathbf{F}}. \tag{13.16}$$

One can use $2\overline{\lambda}$, $3\overline{\lambda}$ or more generally $q\overline{\lambda}$ as a benchmark to determine a set of eigenvalues for analysis. Suppose that

$$\lambda_1 \geqslant \cdots \geqslant \lambda_k \geqslant q\overline{\lambda} > \lambda_{k+1} \geqslant \cdots \lambda_n. \tag{13.17}$$

Recall from (13.7) that the individual measures can be expressed in terms of their coordinates in each eigenvector, namely

$$C_j = \frac{\partial^2 f}{\partial \omega_j^2}(\boldsymbol{\omega}_0) = \sum_{i=1}^{n} (\lambda_i a_{ji}^2). \tag{13.18}$$

If the choice in (13.17) determines that $\mathbf{v}_1$ down to $\mathbf{v}_k$ are relevant, then we call this set of eigenvectors the *pre-normalized q-influential*. We then reduce C_j to

$$C_j(q) = \sum_{i=1}^{k} (\lambda_i a_{ji}^2). \tag{13.19}$$

From this perspective, we consider that the individual influence measure C_j of a perturbation parameter is its *aggregate contribution* to all pre-normialized $q-$influential eigenvectors. If we decide that only the top eigenvector $\mathbf{v}_{\max}$ is relevant, then we reduce our analysis of C_j to an analysis of $\lambda_1 a_{j1}^2$, that is dictated by the j-th coordinate of the maximum eigenvector $\mathbf{v}_{\max}$. In this case, Theorems 13.3.1 and 13.2.1 are applicable.

From (13.19), we have

$$\sum_{j=1}^{n} C_j(q) = \sum_{j=1}^{n} \left(\sum_{i=1}^{k} \lambda_i a_{ij}^2 \right) = \sum_{i=1}^{k} \lambda_i \left(\sum_{j=1}^{n} a_{ij}^2 \right) = \sum_{i=1}^{k} \lambda_i .$$

When $q = 0$,

$$\sum_{j=1}^{n} C_j(0) = \sum_{i=1}^{n} \lambda_i = \mathrm{Tr}\ \ddot{\mathbf{F}}.$$

This suggests that Tr $\ddot{\mathbf{F}}$, or the mean curvature $\overline{\lambda}$ (13.16), can be used to develop a benchmark that judges the magnitude of C_j. Although C_j is bounded by $\lambda_{\max}$ (see (13.1)), $\lambda_{\max}$ can take an arbitrary value and hence, hinders the interpretation of the magnitude of C_j. We address this difficulty by constructing a measure that is equivalent to C_j, but assumes values ranging from 0 to 1. Details will be given in the next chapter.

13.5 Conclusions

In his pioneering work, Cook [?] proposed to use the eigenvector $\mathbf{v}_{\max}$ of the maximum eigenvalue of the Hessian of the log-likelihood to assess local influence. The perturbation parameters that make a significant contribution to the coordinates of $\mathbf{v}_{\max}$ are considered influential. Thus, there is the common perception that $\mathbf{v}_{\max}$ is also a tool for measuring or identifying 'joint influence' [?], notwithstanding that the concepts of individual influence and joint influence were not clearly defined or distinguished. This is in great contrast to the well developed concept of Cook's distance for individual and group deletions, in which the interpretations and limitations of D_i and D_j versus D_{ij} are well understood.

In the discussions within Section 12.1, we establish C_j as a measure of individual influence. In Section 12.5, we establish the maximum eigenvalue $\lambda_{\max}$ as a measure of overall joint influence.

We develop the local influence measures $\{C_j, J_{ij}, \ldots, J_{i_1,\ldots,i_k}, \ldots, J_{1,\ldots,n}\}$ for basic directions $\mathbf{e}_1, \ldots, \mathbf{e}_n$, which can then be incorporated into a systematic framework similar to that for deletion-type Cook's distance measures $\{D_j, D_{ij}, D_{ijk}, \ldots\}$ to establish the relationships among local influence measures. The framework, together with the relationships, enable the comprehensive understanding and rigorous articulation of many prevailing perceptions and insights regarding local influence approaches.

For example, when the magnitude of the maximum eigenvalue $\lambda_{\max}$ dominates other eigenvalues, Theorem 13.2.1 demonstrates that the roles of $C_j, j = 1, \ldots, n$ and the coordinates of $\mathbf{v}_{\max}$ are interchangeable. Under the same assumption, we also find that joint influences are approximated by sums of individual influences. In particular, the overall joint influence is the sum of all individual influences. This observation brings us back to Cook's original insight of working on $\mathbf{v}_{\max}$ to examine whether the j-th perturbation parameter is relatively influential, which explains the perception that working on $\mathbf{v}_{\max}$ enables the simultaneous handling of the perturbation of all perturbation parameters and their joint influence [?].

In the last section, we further explain that the individual influence measure C_j is a sum of its contribution to various eigenvectors, not just the eigenvector $\mathbf{v}_{\max}$. We also argue that one must determine whether one should include additional eigenvectors in one's analysis and the mean curvature provides a reasonable benchmark for deciding how many additional eigenvectors to include.

Chapter 14

Conformal Modifications

In the previous chapters, we demonstrate how to use normal curvature in influence analysis prompted by local perturbation. In this chapter, we fine tune these ideas to address additional needs.

In general, normal curvature can take on any value and is not invariant under a uniform change of scale. As Section 13.4 states while C_j is bounded by $\lambda_{\max}$, the interpretation of C_j and other normal curvatures is hindered by the fact that $\lambda_{\max}$ can take arbitrary values. Despite the fact that in many applications the lack of an objective benchmark does not cause significant difficulties, an objective and yet flexible benchmark is surely welcome and can enhance the applicability of the method.

To facilitate the application and interpretation of the local influence approach, in this chapter we develop some methods to address such difficulties. Specifically, we derive measures that are analogous to Cook's directional curvature for which benchmarks can be established to judge largeness.

14.1 Modification and invariance

We have been considering the perturbation parameters in the space $\mathcal{U}$, which is a subset of the n-dimensional space $\mathbb{R}^n$. There are occasions when one expects that the outcome of an influence analysis will not be affected by a certain type of modification of the perturbation scheme [?].

We conceive of such modifications of perturbation scheme as a map from a space $\mathcal{V}$ with the same number of parameters to $\mathcal{U}$, such that it is has a well-defined inverse $\gamma : \mathcal{V} \to \mathcal{U}$. Mathematically, it is a change of coordinates (See Section 5.1). Although the original influence analysis is based on the log-likelihood displacement function f defined on the parameter space $\mathcal{U}$, a modification of the perturbation scheme yields a second influence analysis based on the modified log-likelihood displacement function defined on the space $\mathcal{V}$ and based on the composite function $g = f \circ \gamma$.

For example, in linear regression analysis, when $\boldsymbol{\omega} = (\omega_1, \ldots, \omega_n)^T$ is the vector of case-weights for a sample of size n and $\mathbf{t} = (t_1, \ldots, t_n)^T$ is the new parameter set, one can consider modifying the perturbation scheme by

$$\omega_i = \frac{1}{2}(1 + t_i), \quad t_i = 2\omega_i - 1.$$

The effect of such a modification was studied in [?]. This example can be generalized to a modification determined by a function $\tau(z)$ of one-variable. Then the modification is given by

$$t_i = \tau(\omega_i). \tag{14.1}$$

Whenever one generates a modification of perturbation schemes, one must understand how the outcome of the influence analysis is affected. In most cases, it is even more important to capture an influence analysis measure that is unchanged or invariant when the perturbation scheme is modified within certain constraints.

Because we consider local perturbation, the differential of the modification γ at the unperturbed point $\boldsymbol{\omega}_0$ becomes a key element for analysis. Consider the modification given in (14.1) for case-weight perturbation. In this case, $\boldsymbol{\omega}_0 = (1, \ldots, 1)^T$. If $\tau(1) = a$, then $\gamma(a, \ldots, a) = \boldsymbol{\omega}_0$ and

$$d\gamma_{|z=1} = \begin{pmatrix} \frac{d\tau}{dz}(1) & 0 & \cdots & 0 \\ 0 & \frac{d\tau}{dz}(1) & \cdots & \vdots \\ \vdots & \vdots & & \vdots \\ 0 & \cdots & 0 & \frac{d\tau}{dz}(1) \end{pmatrix} = \frac{d\tau}{dz}_{|z=1} \mathbf{I}. \tag{14.2}$$

This type of modification is an example of the non-linear conformal transformation as introduced in Section 5.2.

14.2 Invariant measures

Suppose that a perturbation scheme is given by a log-likelihood displacement function f defined on a parameter space $\mathcal{U} \in \mathbb{R}^n$ with $\boldsymbol{\omega}_0$ as its unperturbed point. Now, suppose that a map $\gamma : \mathcal{V} \to \mathcal{U}$ has an inverse $\gamma^{-1} : \mathcal{U} \to \mathcal{V}$, and $\mathbf{t}_0 \in \mathcal{V}$, such that $\gamma(\mathbf{t}_0) = \boldsymbol{\omega}_0$. Then the composite function $g = f \circ \gamma$ is a modification of f. This is the log-likelihood displacement function of a perturbation whose parameters are in the set $\mathcal{V}$ with $\mathbf{t}_0$ as its unperturbed point. We consider the modification γ to be conformal if the differential of the related map γ at the unperturbed point $\boldsymbol{\omega}_0$ is a conformal matrix, which means that there exists a constant $c > 0$ such that

$$d\gamma^T(\mathbf{t}_0)d\gamma(\mathbf{t}_0) = c^2\mathbf{I}. \tag{14.3}$$

A modification is a *homothety* if

$$d\gamma(\mathbf{t}_0) = c\mathbf{I}. \tag{14.4}$$

A homothetic modification is also conformal. We seek measures that are invariant of under conformal modifications.

Mathematical preparations in this regard are already given in Chapter 5. We recall the key observations here. The unperturbed point $\boldsymbol{\omega}_0$ of the log-likelihood displacement function f is a minimum point of the function f. Therefore, the computation in Section 5.3 is applicable. In particular, consider the quantity, as given in (5.14), for the function f at $\boldsymbol{\omega}_0$,

$$B_{\boldsymbol{\ell}} = \frac{1}{\sqrt{\operatorname{Tr}\ddot{\mathbf{F}}^2}}\frac{\boldsymbol{\ell}^T\ddot{\mathbf{F}}\boldsymbol{\ell}}{\boldsymbol{\ell}^T\boldsymbol{\ell}} = \frac{1}{\sqrt{\operatorname{Tr}\ddot{\mathbf{F}}^2}}C_{\boldsymbol{\ell}}. \tag{14.5}$$

This is invariant with respect to a conformal modification. We call it the *conformal normal curvature* at a critical point $\boldsymbol{\omega}_0$ of a graph in the direction $\boldsymbol{\ell}$.

Based on the computation in Section 12.2, we further deduce the conformal invariant curvature in terms of the Hessian of the log-likelihood function. In terms of the notations given from (12.10) to (12.13), the conformal normal curvature of the graph over the point $\boldsymbol{\omega}_0$ in a direction $\boldsymbol{\ell}$ with $\boldsymbol{\ell}^T\boldsymbol{\ell} = 1$ is

$$B_{\boldsymbol{\ell}} = \frac{\boldsymbol{\ell}^T\ddot{\mathbf{F}}\boldsymbol{\ell}}{\sqrt{\operatorname{Tr}\ddot{\mathbf{F}}^2}} = -\frac{\boldsymbol{\ell}^T\boldsymbol{\Delta}^T\ddot{\mathbf{L}}^{-1}\boldsymbol{\Delta}\boldsymbol{\ell}}{\sqrt{\operatorname{Tr}(\boldsymbol{\Delta}^T\ddot{\mathbf{L}}^{-1}\boldsymbol{\Delta})^2}}. \tag{14.6}$$

It is clear from (14.5) that the computation of $B_{\boldsymbol{\ell}}$ requires almost no additional effort compared to the computation of $C_{\boldsymbol{\ell}}$. Moreover, the conformal normal curvature enjoys numerous properties, such as assuming values in the unit interval.

14.3 Benchmarks

In Section 12.5, we conclude that the maximum eigenvalue for the Hessian $\ddot{\mathbf{F}}$ at the unperturbed point $\boldsymbol{\omega}_0$ is a measure of overall joint influence. In Section 13.2, we find that sometimes the eigenvector $\mathbf{v}_{\max}$ is a useful tool for assessing individual influence. However, in Section 13.4, we argue that one must understand whether other eigenvalues and their related eigenvectors play a role in an analysis of joint influences. Therefore, one must have an effective method for benchmarking eigenvalues. In fact, in Section 13.4, we indicate how to use mean curvature as a benchmark for normal curvatures. In this section, we work with the conformal normal curvature $B_{\boldsymbol{\ell}}$.

We continue to use the notations as given in Section 13.2. In particular, $\{\mathbf{v}_1, \ldots, \mathbf{v}_n\}$ is an orthonormal basis of eigenvectors for the matrix $\ddot{\mathbf{F}}$. The corresponding eigenvalues are λ_i for $i = 1, \ldots, n$. In terms of the standard basis $\{\mathbf{e}_1, \ldots, \mathbf{e}_n\}$, the coordinates of the eigenvectors $\mathbf{v}_i$ are given in the matrix $\mathbf{A}$ of (13.4).

Take any $\boldsymbol{\ell}$ as a linear combination of eigenvectors $\boldsymbol{\ell} = \sum_i \ell_i \mathbf{v}_i$, then $\boldsymbol{\ell}^T \ddot{\mathbf{F}} \boldsymbol{\ell} = \sum_i \lambda_i \ell_i^2$. It follows that

$$|B_{\boldsymbol{\ell}}| = \frac{\left|\sum_i \lambda_i \ell_i^2\right|}{\left(\sum_i \ell_i^2\right) \sqrt{\sum_i \lambda_i^2}}.$$

According to the Schwartz Inequality, $|\sum_i \lambda_i \ell_i^2| \leqslant \sqrt{\sum_i \lambda_i^2} \sqrt{\sum_i \ell_i^4}$. Because $(\sum_i \ell_i^2)^2 = \sum_i \ell_i^4 + 2\sum_{i<j} \ell_i^2 \ell_j^2$,

$$|B_{\boldsymbol{\ell}}| \leqslant \frac{\sqrt{\sum_i \ell_i^4}}{\sqrt{\sum_i \ell_i^4 + 2\sum_{i<j} \ell_i^2 \ell_j^2}} \leqslant 1.$$

We summarize the above computation in the next observation.

Theorem 14.3.1 *For any direction $\boldsymbol{\ell}$, the conformal normal curvature $B_{\boldsymbol{\ell}}$ satisfies the condition that $0 \leqslant |B_{\boldsymbol{\ell}}| \leqslant 1$.*

This theorem tells us that $B_{\boldsymbol{\ell}}$ is a normalized measure and thus it becomes easier to interpret its magnitude. Furthermore,

$$B_{\mathbf{v}_i} = \frac{\mathbf{v}_i^T \ddot{\mathbf{F}} \mathbf{v}_i}{\sqrt{\operatorname{Tr} \ddot{\mathbf{F}}}} = \frac{\lambda_i}{\left(\sum_{k=1}^{n} \lambda_k^2\right)^{1/2}}. \tag{14.7}$$

This is the normalized eigenvalue, which we denote by λ_i^*. It is an eigenvalue of the matrix of the bilinear form of $B_{\boldsymbol{\ell}}$ in (14.5). It can now be seen that

$$\sum_{i=1}^{n} B_{\mathbf{v}_i}^2 = 1. \tag{14.8}$$

Given identity (14.8) above, we know that if the conformal normal curvature for all eigenvectors is identical, they are all equal to $\frac{1}{\sqrt{n}}$. Given $\frac{1}{\sqrt{n}}$ as a benchmark for eigenvalues, we can make objective decisions. For instance, a principal direction (i.e., eigenvector) $\mathbf{v}$ has a below average influence if $|B_{\mathbf{v}}| < \frac{1}{\sqrt{n}}$. We could use $\frac{2}{\sqrt{n}}$ or $\frac{3}{\sqrt{n}}$ as our benchmark to determine outlying behavior.

For example, for the Data Set B given in Table 12.2, the first two normalized eigenvalues are, respectively,

$$\lambda_1^* = 0.9470, \quad \text{and} \quad \lambda_2^* = 0.3213.$$

Because $\frac{1}{\sqrt{n}} = 0.2887$, the second eigenvector is regarded as influential with respect to $\frac{1}{\sqrt{n}}$; but not influential with respect to $\frac{2}{\sqrt{n}}$.

In general, for any integer $1 \leqslant q < \sqrt{n}$, an eigenvector $\mathbf{v}$ is called *q-influential* if $|B_{\mathbf{v}}| \geqslant \frac{q}{\sqrt{n}}$[?]. Equivalently, the direction $\mathbf{v}_i$ is q-influential if $\lambda_i^* \geqslant \frac{q}{\sqrt{n}}$. There is an integer k, such that

$$\lambda_1^* \geqslant \cdots \geqslant \lambda_k^* \geqslant \frac{q}{\sqrt{n}} > \lambda_{k+1}^* \geqslant \cdots \geqslant \lambda_n^*. \tag{14.9}$$

If $k = 1$, there is a single dominant eigenvalue and principal direction $\mathbf{v}_1$. In this case, any analysis of the relation between individual influence and joint influence is reduced to our discussion in Section 13.2 because there is a single dominant eigenvalue.

We could apply the same idea to use the average of $B_{\mathbf{e}_j}$ as a benchmark for individual influence $B_{\mathbf{e}_j}$. Because

$$\sum_{j=1}^{n} B_{\mathbf{e}_j} = \frac{1}{\sqrt{\operatorname{Tr}\ddot{\mathbf{F}}^2}} \sum_j C_{\mathbf{e}_j} = \frac{\operatorname{Tr}\ddot{\mathbf{F}}}{\sqrt{\operatorname{Tr}\ddot{\mathbf{F}}^2}},$$

and the mean curvature $\overline{\lambda}$ is precisely

$$\overline{\lambda} = \frac{1}{n}\operatorname{Tr}\ddot{\mathbf{F}} = \frac{1}{n}(\lambda_1 + \cdots + \lambda_n),$$

According to (5.11) $\operatorname{Tr}\ddot{\mathbf{F}}^2 = \lambda_1^2 + \cdots + \lambda_n^2$, the benchmark for conformal normal curvature is

$$b = \frac{1}{n}\frac{\operatorname{Tr}\ddot{\mathbf{F}}}{\sqrt{\operatorname{Tr}\ddot{\mathbf{F}}^2}} = \frac{\lambda_1 + \cdots + \lambda_n}{n\sqrt{\lambda_1^2 + \cdots + \lambda_n^2}}. \tag{14.10}$$

14.4 Individual's contribution to joint influence — re-visited

When we explore the relation between individual influence and joint influence in Section 13.2, we learn that it is important to judge which principal directions (i.e., eigenvectors) are influential, and which individual parameters are influential, with the understanding that such judgments are often related.

Continue to use the notation in Section 13.2, we have the relation between the eigenvectors and the standard basis as given in equation (13.4).

$$\mathbf{v}_i = \sum_{j=1}^{n} a_{ji}\mathbf{e}_j, \quad \text{and} \quad \mathbf{e}_j = \sum_{i=1}^{n} a_{ji}\mathbf{v}_i.$$

In the last section, we indicated why $q/\sqrt{n}$ serves as a benchmark for selecting q-influential eigenvectors. To detect the collection of perturbation parameters making simultaneous contributions to these directions, we define the *aggregate contribution* of the j-th perturbation vector to be

$$B_j(q) = \sum_{i=1}^{k} \lambda_i^* a_{ji}^2, \tag{14.11}$$

where k is determined by (14.9). Because the matrix $\mathbf{A} = (a_{ij})$ is an orthogonal matrix,

$$\sum_{j=1}^{n} B_j(q) = \sum_{j=1}^{n}\left(\sum_{i=1}^{k} \lambda_i^* a_{ij}^2\right) = \sum_{i=1}^{k} \lambda_i^* \left(\sum_{j=1}^{n} a_{ij}^2\right) = \sum_{i=1}^{k} \lambda_i^*.$$

If the aggregate contribution of each basic perturbation vector is the same, then each is equal to

$$b(q) = \frac{1}{n}\sum_{i=1}^{k} \lambda_i^* = \frac{1}{n}\frac{\sum_{i=1}^{k}\lambda_i}{\sqrt{\sum_{i=1}^{n}\lambda_i^2}}. \tag{14.12}$$

Therefore, we refer to $b(p)$ as a benchmark when determining the significance of the contribution of the individual basic perturbation vectors.

When $q = 0$, according to (14.10), $b(0) = b$. Based on (14.11),

$$B_j(0) = \sum_{i=1}^{n} \lambda_i^* a_{ji}^2 = \frac{1}{\sqrt{\sum_{i=1}^{n}\lambda_i^2}}\sum_{i=1}^{n} \lambda_i a_{ji}^2. \tag{14.13}$$

According to (13.7), it is equal to

$$\frac{1}{\sqrt{\sum_{i=1}^{n}\lambda_i^2}} C_j.$$

Given (14.5), $B_j(0) = B_j$.

Therefore, treating B_j as measures of individual influences, the above iteration from $B_j(p)$ for large p to $B(0)$ means that the aggregate contribution of a perturbation parameter to all eigenvector directions is precisely its individual influence.

In addition, when all perturbation parameters are equally influential, it means that $B_j = B_k$ for all j, k. In this case, each is equal to the b as given in equation (14.10). This can be used to establish objective benchmarks that facilitate the judgment of whether individual perturbations are influential. For instance, for the Data Set C given in Table 13.1, the index plot of the conformal normal curvature along with the values b and $2b$ are given in Figure 14.1.

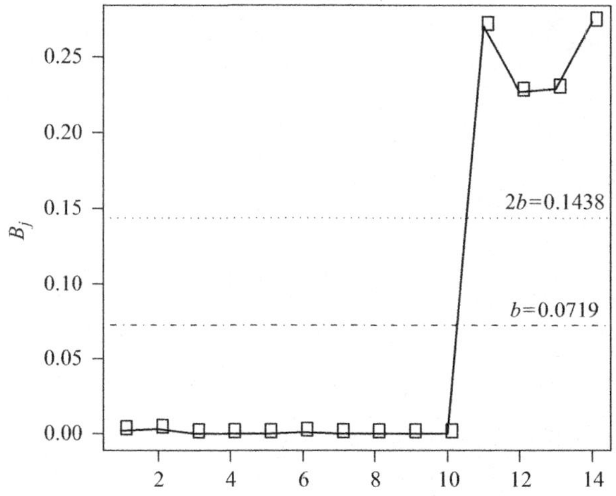

Fig. 14.1 Index plot of B_j for data set C

This demonstrates that the local influence of Cases 11, 12, 13 and 14 are substantial.

Appendix A Rank of Hat Matrix

Much of the discussion in Part III of this book addresses the eigenvalues and eigenvectors of the Hessian $\ddot{\mathbf{F}}$ of the log-likelihood function of a perturbation scheme at the unperturbed point $\boldsymbol{\omega}_0$. We argue in Section 13.4 that one must sometimes analyze the contribution of various perturbation parameters to the second eigenvalues or beyond. Therefore, it is necessary to have an understanding of the number of non-zero eigenvalues for $\ddot{\mathbf{F}}$. The answer, of course, depends on the statistical model and the perturbation scheme. If all non-zero eigenvalues of $\ddot{\mathbf{F}}$ are different, then the *rank* of the matrix $\ddot{\mathbf{F}}$ is precisely the number of non-zero eigenvalues.

We now focus on the key examples in this book. It is shown in Section 12.3 that in a linear regression model with case-weight perturbation scheme, according to (12.21),

$$\ddot{\mathbf{F}} = \frac{2}{\sigma^2}\mathbf{R}^T\mathbf{X}(\mathbf{X}^T\mathbf{X})^{-1}\mathbf{X}^T\mathbf{R} = \frac{2}{\sigma^2}\mathbf{RHR}, \tag{A.1}$$

where $\mathbf{H}$ is the hat matrix $\mathbf{H} = \mathbf{X}(\mathbf{X}^T\mathbf{X})^{-1}\mathbf{X}^T$ and $\mathbf{R} = \mathrm{Diag}(r_1, \ldots, r_n)$. In general, when none of the residuals is equal to zero, the number of non-zero eigenvalues of $\ddot{\mathbf{F}}$ is precisely the number of non-zero eigenvalues of the hat matrix $\mathbf{H}$. As seen in Section 2.6, the hat matrix $\mathbf{H}$ is a projection matrix that projects from $\mathbb{R}^n$ on the the subspace spanned by the column vectors in the matrix $\mathbf{X}$. Therefore, $\mathbf{H}$ has only "0" and "1" as its eigenvectors, and the number of non-zero eigenvectors is exactly the number of columns in $\mathbf{X}$.

In a multiple regression model such as these given in (10.1), the number of columns is $k+1$, where k is the number of independent variables. In particular, for a simple linear regression model with a single variable, the number of non-zero eigenvalues for the hat matrix is equal to "2".

Appendix B Ricci Curvature

In Section 12.4, when we searched for joint influence measures, we analyzed sectional curvature and then dismissed it as inadequate. One might ask if there are other common geometric quantities that might serve our purpose. One important curvature concept in differential geometry is called Ricci curvature. In this section, we develop this concept and review its role in our context.

Recall that for each pair of basic vectors $\mathbf{e}_i$ and $\mathbf{e}_j$, the sectional curvature $\mathrm{Sec}(\mathbf{e}_i, \mathbf{e}_j)$ of the graph of a function f at a critical point is equal to $\det \ddot{\mathbf{F}}_{(ij)}$, where the 2×2-matrix is given in (4.22). Therefore, for each $\mathbf{e}_i$ the quantity

$$\sum_{j \neq i} \mathrm{Sec}(\mathbf{e}_i, \mathbf{e}_j) = \sum_{j=1}^{n} \left(\left(\frac{\partial^2 f}{\partial \omega_i^2} \right) \left(\frac{\partial^2 f}{\partial \omega_j^2} \right) - \left(\frac{\partial^2 f}{\partial \omega_i \partial \omega_j} \right)^2 \right) \tag{B.1}$$

represents the total sectional curvature with respect to all possible coordinate planes in which $\mathbf{e}_i$ is a fixed choice of basic vectors. It is also equal to

$$\begin{aligned} &\left(\frac{\partial^2 f}{\partial \omega_i^2} \right) \sum_{j=1}^{n} \left(\frac{\partial^2 f}{\partial \omega_j^2} \right) - \sum_{j=1}^{n} \left(\frac{\partial^2 f}{\partial \omega_i \partial \omega_j} \right)^2 \\ &= \mathbf{e}_i^T \left((\mathrm{Tr}\, \ddot{\mathbf{F}}) \ddot{\mathbf{F}} - \ddot{\mathbf{F}}^2 \right) \mathbf{e}_i . \end{aligned} \tag{B.2}$$

Given two tangent vectors, $\mathbf{v}$ and $\mathbf{w}$, for the graph of the function f at a critical point, the Ricci curvature of the linear span of $\mathbf{v}$ and $\mathbf{w}$ is defined by

$$\mathrm{Ric}(\mathbf{v}, \mathbf{w}) = \mathbf{v}^T ((\mathrm{Tr}\, \ddot{\mathbf{F}}) \ddot{\mathbf{F}} - \ddot{\mathbf{F}}^2) \mathbf{w}. \tag{B.3}$$

The symmetric matrix

$$\mathrm{Ric} = (\mathrm{Tr}\, \ddot{\mathbf{F}}) \ddot{\mathbf{F}} - \ddot{\mathbf{F}}^2 \tag{B.4}$$

is the *Ricci curvature* of the graph at points where $\nabla_f = \mathbf{0}$.

If $\mathbf{v}_i$ is a unit eigenvector of $\ddot{\mathbf{F}}$ with eigenvalue λ_i, according to (B.4)

$$\begin{aligned}
&((\operatorname{Tr}\ddot{\mathbf{F}})\ddot{\mathbf{F}} - \ddot{\mathbf{F}}^2)\mathbf{v}_i \\
&= \left(\sum_{j=1}^{n} \lambda_j\right) \lambda_1 \mathbf{v}_i - \lambda_i^2 \mathbf{v}_i \\
&= \lambda_i \left(\sum_{j\neq i} \lambda_j\right) \mathbf{v}_i.
\end{aligned}$$

Therefore, the matrix Ric and the Hessian matrix $\ddot{\mathbf{F}}$ share the same set of eigenvectors. For $\mathbf{v}_i$, the eigenvalue with respect to Ric is $\lambda_i(\sum_{j\neq i} \lambda_j)$.

If all of the eigenvalues of $\ddot{\mathbf{F}}$ are non-negative, then

$$\begin{aligned}
&\operatorname{Ric}(\mathbf{v}_i, \mathbf{v}_i) - \operatorname{Ric}(\mathbf{v}_j, \mathbf{v}_j) \\
&= (\lambda_i - \lambda_j) \sum_{k\neq i,j} \lambda_k.
\end{aligned}$$

It follows that the order of the eigenvalues of Ric is identical to the order of the eigenvalues of $\ddot{\mathbf{F}}$. In particular, the maximum eigenvalue of Ric is realized in the direction of $\mathbf{v}_{\max}$, which is equal to

$$\lambda_1 (\lambda_2 + \cdots + \lambda_n) = \lambda_{\max} (\lambda_2 + \cdots + \lambda_n). \tag{B.5}$$

According to Proposition 1.2.3, the maximum eigenvalue of Ric is also the maximum value of $\operatorname{Ric}(\mathbf{v}, \mathbf{v})$ among all possible unit vectors $\mathbf{v}$.

Therefore, the Ricci curvature of the graph of the function f at a critical point of the function does not carry more information than that of the curvature function C given in Section 4.11.

Appendix C Cook's Distance – Deleting Two Data Points

In this appendix, we continue our computation in Section 10.7.2, and derive the Cook's distance of deleting two data points.

When one deletes two cases, e.g., the i-th and the j-th cases, then one sets $\boldsymbol{\omega} = (1, \ldots, 1, 0, 1, \ldots, 1, 0, 1, \ldots, 1)$ where 0 occurs in the i-th and the j-th slots. One can repeat the computation of Section 10.7.2 to determine the Cooks' distance D_{ij} defined in (10.25). See also [?]. Therefore,

$$\mathbf{H}_{\boldsymbol{\omega}} = \mathbf{X}(\mathbf{X}^T\boldsymbol{\Omega}\mathbf{X})^{-1}\mathbf{X}^T\boldsymbol{\Omega},$$

where $\boldsymbol{\Omega} = \mathbf{I} - \mathbf{e}_i\mathbf{e}_i^T - \mathbf{e}_j\mathbf{e}_j^T$. To calculate $(\mathbf{X}^T\boldsymbol{\Omega}\mathbf{X})^{-1}$, we apply (10.39) twice to get

$$\begin{aligned}
&(\mathbf{X}^T\boldsymbol{\Omega}\mathbf{X})^{-1}\\
&= (\mathbf{X}^T\mathbf{X})^{-1} + \frac{1-h_{ii}}{\Delta}(\mathbf{X}^T\mathbf{X})^{-1}\mathbf{X}_j\mathbf{X}_j^T(\mathbf{X}^T\mathbf{X})^{-1}\\
&\quad + \frac{1-h_{jj}}{\Delta}(\mathbf{X}^T\mathbf{X})^{-1}\mathbf{X}_i\mathbf{X}_i^T(\mathbf{X}^T\mathbf{X})^{-1}\\
&\quad + \frac{h_{ij}}{\Delta}(\mathbf{X}^T\mathbf{X})^{-1}\mathbf{X}_j\mathbf{X}_i^T(\mathbf{X}^T\mathbf{X})^{-1} + \frac{h_{ji}}{\Delta}(\mathbf{X}^T\mathbf{X})^{-1}\mathbf{X}_i\mathbf{X}_j^T(\mathbf{X}^T\mathbf{X})^{-1},
\end{aligned}$$

where

$$\Delta = \det\left(\mathbf{I} - \begin{pmatrix} h_{ii} & h_{ij} \\ h_{ji} & h_{jj} \end{pmatrix}\right) = 1 - (h_{ii} + h_{jj}) + (h_{ii}h_{jj} - h_{ij}^2).$$

Next,

$$\mathbf{X}^T\boldsymbol{\Omega}\mathbf{Y} = \mathbf{X}^T(\mathbf{I} - \mathbf{e}_i\mathbf{e}_i^T - \mathbf{e}_j\mathbf{e}_j^T)\mathbf{Y} = \mathbf{X}^T\mathbf{Y} - \mathbf{X}_iY_i - \mathbf{X}_jY_j.$$

Because $\mathbf{H}\mathbf{e}_j = \mathbf{X}(\mathbf{X}^T\mathbf{X})^{-1}\mathbf{X}_j$, we have

$$
\begin{aligned}
&\mathbf{X}\widehat{\boldsymbol{\beta}}_{\boldsymbol{\omega}} - \mathbf{X}\widehat{\boldsymbol{\beta}} \\
&= \frac{1-h_{ii}}{\Delta}\mathbf{He}_j(\mathbf{He}_j)^T\mathbf{Y} + \frac{1-h_{jj}}{\Delta}\mathbf{He}_i(\mathbf{He}_i)^T\mathbf{Y} + \frac{h_{ij}}{\Delta}\mathbf{He}_j(\mathbf{He}_i)^T\mathbf{Y} \\
&\quad + \frac{h_{ji}}{\Delta}\mathbf{He}_i(\mathbf{He}_j)^T\mathbf{Y} \\
&\quad - \frac{1-h_{jj}}{\Delta}(\mathbf{He}_i)Y_i - \frac{1-h_{ii}}{\Delta}(\mathbf{He}_j)Y_j - \frac{h_{ij}}{\Delta}(\mathbf{He}_j)Y_i \\
&\quad - \frac{h_{ij}}{\Delta}(\mathbf{He}_i)Y_j.
\end{aligned}
$$

Given

$$
r_i = Y_i - \mathbf{X}_i^T(\mathbf{X}^T\mathbf{X})^{-1}\mathbf{X}^T\mathbf{Y} = Y_i - \mathbf{X}_i^T\widehat{\boldsymbol{\beta}} = Y_i - (\mathbf{He}_i)^T\mathbf{Y}, \tag{C.1}
$$

$$
\begin{aligned}
&\mathbf{X}\widehat{\boldsymbol{\beta}}_{\boldsymbol{\omega}} - \mathbf{X}\widehat{\boldsymbol{\beta}} \\
&= -\frac{(1-h_{ii})r_j}{\Delta}\mathbf{He}_j - \frac{(1-h_{jj})r_i}{\Delta}\mathbf{He}_i - \frac{h_{ij}r_i}{\Delta}\mathbf{He}_j - \frac{h_{ij}r_j}{\Delta}\mathbf{He}_i \\
&= -\frac{(1-h_{ii})r_j + h_{ij}r_i}{\Delta}\mathbf{He}_j - \frac{(1-h_{jj})r_i + h_{ij}r_j}{\Delta}\mathbf{He}_i.
\end{aligned}
$$

Given $(\mathbf{He}_i)^T(\mathbf{He}_j) = \mathbf{e}_i^T\mathbf{HHe}_j = \mathbf{e}_i^T\mathbf{He}_j = h_{ij}$ for all i, j,

$$
\begin{aligned}
&|\mathbf{X}\widehat{\boldsymbol{\beta}}_{\boldsymbol{\omega}} - \mathbf{X}\widehat{\boldsymbol{\beta}}|^2 \\
&= \frac{1}{\Delta^2}\left(((1-h_{ii})r_j + h_{ij}r_i)^2 h_{jj}\right. \\
&\quad + 2((1-h_{ii})r_j + h_{ij}r_i)((1-h_{jj})r_i + h_{ij}r_j)h_{ij} \\
&\quad \left. + ((1-h_{jj})r_i + h_{ij}r_j)^2 h_{ii}\right).
\end{aligned}
$$

Then, from (10.34),

$$
\begin{aligned}
D_{ij} = &\frac{1}{(k+1)\widehat{\sigma}^2\Delta^2}\left(((1-h_{ii})r_j + h_{ij}r_i)^2 h_{jj}\right. \\
&+ 2((1-h_{ii})r_j + h_{ij}r_i)((1-h_{jj})r_i + h_{ij}r_j)h_{ij} \\
&\left. + ((1-h_{jj})r_i + h_{ij}r_j)^2 h_{ii}\right).
\end{aligned} \tag{C.2}
$$

Bibliography

[1] A. C. Atkinson, *Plots, Transformations and Regression*, Oxford Statistical Science Series 1, Oxford University Press, Oxford, (1985).

[2] M. Barros, M. Galea, M. González & V. Leiva. Influence diagnostics in the tobit censored response model, *Statistical Methods & Applications*, 19, (2010), 379–397.

[3] R.J. Beckman, C.J. Nachtsheim & R. D. Cook, Diagnostics for mixed-model analysis of variance, *Technometrics*, 29, (1987), 413 – 426.

[4] N.G. Cadigan, Local influence diagnostics for quasi-likelihood and log-normal estimates of a biological reference point from some fish stock and recruitment models, *Biometrics,* 62, (2006), 713–720.

[5] R.D. Cook, Detection of influential observations in linear regression, *Technometrics*, 19, (1977), 15–18.

[6] R.D. Cook, Assessment of local influence (with discussion), *Journal of the Royal Statistical Society Series B*, 48, (1986), 133–169.

[7] R.D. Cook, D. Pena & S. Weisberg, The likelihood displacement: A unifying principle for influence measures, *Communications in Statistics: Theory and Methods*, 17, (1988), 623–640.

[8] W.K. Fung & C.W. Kwan, A note on local influence based on normal curvature, *Journal of the Royal Statistical Society, Series B*, 59, (1997), 839–843.

[9] G. Ibacache-Pulgar & G.A. Paula, Local influence for student-t partially linear models, *Computational Statistics and Data Analysis*, 55, (2011), 1462–1478.

[10] R.W. Kennard, A note on the C_p Statistic, *Technometrics*, 13, (1971), 899–900.

[11] V.H. Lachos, T. Angoline & C.A. Abanto-Valle, On estimation and local influence analysis for measurement errors models under heavy-tailed distributions, *Statistics Papers*, 52, (2011), 567–590.

[12] A.J. Lawrance, Regression transformation diagnostics using local influence, *Journal of the American Statistical Association*, 83, (1988), 1067–1072.

[13] A.J. Lawrance, Local and deletion influence. In *Direction in Robust Statistics and Diagnostics, Part* 1 (eds. W. Stahel and S. Weisberg) Wiley, New York, (1991), 141–157.

[14] A.J. Lawrance, Deletion influence and masking in regression, *Journal of the Royal Statistical Society, Series B*, 57, (1995), 181–189.

[15] S.Y. Lee & S.J. Wang, Sensitivity analysis of structural equation models, *Psychometrika*, 61, (1996), 93–108.

[16] E. Lesaffre & G. Verbeke, Local influence in linear mixed models, *Biometrics*, 54, (1998), 570–582.

[17] R.M. Loynes, Discussion on assessment of local influence (by R.D. Cook), *Journal of the Royal Statistical Society, Series B*, 48, (1986), 156–157.

[18] B. Lu & X.Y. Song, Local influence analysis of multivariate probit latent variable models, *Journal of Multivariate Analysis*, 97, (2006), 1783–1798.

[19] L.C. Montenegro, V.H. Lachos. & H. Bolfarine, Local influence analysis for skew-normal linear mixed models, *Communications in Statistics: Theory and Methods*, 38, (2009), 484–496.

[20] F. Osorio, G.A. Paula & M. Galea, Assessment of local influence in elliptical linear models with longitudinal structure, *Computational Statistics & Data Analysis*, 51, (2007), 4354–4368.

[21] G.A. Paula & F.J.A. Cysneiros, Local influence under parameter constraints, *Communications in Statistics: Theory and Methods*, 39, (2010), 1212–1228.

[22] D. Peña & V. Yohai, The detection of influential subsets in linear regression by using an influence matrix, *Journal of the Royal Statistical Society, Series B*, 57, (1995), 145–156.

[23] W.Y. Poon, Identifying influential observations in discriminant analysis, *Statistical Methods in Medical Research: An International Review Journal*, 13, (2004), 291–308.

[24] W.Y. Poon & Y.S. Poon, Conformal normal curvature and assessment of local influence, *Journal of the Royal Statistical Society, Series B*, 61, (1999), 51–61.

[25] W.Y. Poon & Y.S. Poon, Influential observations in the estimation of mean vector and covariance matrix, *British Journal of Mathematical and Statistical Psychology*, 55, (2002), 177–192.

[26] W.Y. Poon, M.L. Tang & S. J. Wang, Influence measures in contingency tables with application in sampling zeros, *Sociological Methods and Research*, 31, (2003), 439–452.

[27] C.M. Russo, G.A. Paula & R. Aoki, Influence diagnostics in nonlinear mixed-effects elliptical models, *Computational Statistics & Data Analysis*, 53, (2009), 4143–4156.

[28] N. Sartori, K. Thomaseth & A. Salvan, Local influence analysis when interfacing toxicokinetic and proportional hazard models, *Statistics in Medicine*, 23, (2004), 2399–2412.

[29] B. Schwarzmann, A connection between local-influence analysis and residual diagnostics, *Technometrics*, 33, (1991), 103–104.

[30] L. Shi, Local influence in principal component analysis, *Biometrika*, 84, (1997), 175–186.

[31] M. Spivak, *Calculus on Manifolds*, Westview Press, Boulder (1971).

[32] R.Y. St Laurent & R.D. Cook, Leverage, local influence and curvature in nonlinear regression, *Biometrika*, 80, (1993), 99–106.

[33] N.S. Tang, B.C. Wei & X.R. Wang, Detection of influential observations in nonlinear reproductive dispersion models, *Australian and New Zealand Journal of Statistics*, 44, (2002), 185–192.

[34] W. Thomas & R.D. Cook, Assessing influence on predictions from generalized linear models, *Technometrics*, 32, (1990), 59–65.

[35] K.L.P. Vasconcellos & L.M.Z. Fernandez, Influence analysis with homogeneous linear restrictions, *Computational Statistics & Data Analysis*, 53, (2009), 3787–3794.

[36] G. Verbeke, G. Molenberghs, H. Thijs, E. Lesaffre & M.G. Kenward, Sensitivity analysis for nonrandom dropout: A local influence approach, *Biometrics*, 57, (2001), 7–14.

[37] L. Xu, W.Y. Poon & S. Y. Lee, Influence analysis for the factor analysis model with ranking data, *British Journal of Mathematical and Statistical Psychology*, 61, (2008), 133–161.

[38] L. Shi & M. Huang, Stepwise local influence analysis, *Computational Statistics & Data Analysis*, 55, (2011), 973–982.

[39] C.B. Zeller, F.V. Labra, V.H. Lachos & N. Balakrishnan, Influence analyses of skew-normal/independent linear mixed models, *Computational Statistics & Data Analysis*, 54, (2010), 1266–1280.

[40] C.B. Zeller, V.H. Lachos & F.E. Vilca-Labra, Local influence analysis for regression models with scale mixtures of skew-normal distributions, *Journal of Applied Statistics*, 38, (2011), 343–368.

[41] J. Zu & K.H. Yuan, Local influence and robust procedures for mediation analysis, *Multivariate Behavioral Research*, 45, (2010) 1–44.

[42] H.T. Zhu & S.Y. Lee, Local influence for incomplete-data models, *Journal of the Royal Statistical Society, Series B*, 63, (2001), 111–126.

[43] H. Zhu, J.G. Ibrahim, S. Lee & H. Zhang, Perturbation selection and influence measures in local influence analysis, *Annals of Statistics*, 35, (2007), 2565–2588.

Index

C_j, 130, 131
F distribution, 71
q-influential, 149, 157
t-distribution, 71

A

asymptotically unbiased, 76, 100

B

basic vectors, 4
basis, 4
Bernoulli distribution, 62
Bernoulli experiment, 62
Bernoulli trial, 62
bilinear form, 5

C

chi-squared distribution, 71
coefficient of determination, 101, 103
coefficient of multiple correlation, 103, 109
coefficient of multiple determination, 109
conditional mathematical expectation, 86
conditional mean, 86
conditional probabilities, 79
conditional probability function, 79
conditional variance, 86
conformal aggregate contribution, 158
conformal map, 20
conformal matrices, 20
conformal normal curvature, 155
conformal transformation, 44
continuous random variable, 51–53, 65
Cook's curvature, 131
Cook's directional curvature, 131
Cook's distance, 111, 113
correlation coefficient, 82
correlation matrix, 91
covariance matrix, 88
critical point, 12
cross products of deviations, 96
cross section, 33
cumulative distribution function, 55

D

determinant, 8
deviation score, 58
deviations from the mean, 96
diagonal, 5
dimension, 6
directional derivative, 35
discrete random variable, 52, 52
dispersion measures, 58
distribution, 113
distribution function, 66

E

eigenvalue, 7
eigenvector, 7
expected value, 56, 60

F

first fundamental form, 34

frequency function, 53

G

gradient vector field, 11
group, 16
Guassian curvature, 39

H

hat matrix, 107, 133
high leverage points, 112

I

i.i.d, 53, 54
idempotent, 24
identically distributed, 53
identically independently distributed, 53
identity matrix, 5
independent, 53
independent variable, 93
infinitesimal perturbation, 124
influence graph, 128
invariant subspaces, 7
inverse, 7
invertible, 7
isometry, 15, 43

J

joint influence, 114
joint marginal distribution, 90
joint probability density function, 80
joint probability function, 77, 90

L

least squares, 94
least squares estimates, 97
least squares estimators, 96, 97
least squares line, 97
least squares prediction line, 97
least squares regression line, 95, 97
leverage, 111, 111
likelihood displacement, 127
likelihood equations, 61
likelihood function, 61, 72
linear, 4
linear combination, 3
linear regression model, 52
linear span, 3
linearly independent, 3
local individual influence, 130
local influence analysis, 124
local joint influence, 138, 142, 150
local perturbation, 124
location measures, 57
log-likelihood, 61
log-likelihood displacement, 127

M

Mahalanobis distance, 92
marginal probability density function, 80
marginal probability functions, 78
masking effects, 114
mathematical expectation, 54, 56, 60, 60, 66
maximum likelihood, 61
maximum likelihood estimate, 61, 72
maximum likelihood estimator, 62
maximum likelihood method, 61
mean, 94
mean curvature, 38, 149
mean of absolute deviations, 58
mean squared error, 100
multiple correlation coefficient, 103
multiple regression, 93
multiple regression model, 105

N

natural gap, 113
normal curvature, 37
normal curvature function, 37

normal distribution, 51
normal section, 32
normal vector, 7
normal vectors of the graph, 30
normalized, 20
null perturbation, 126

O

orthogonal basis, 20
orthogonal complement, 7, 21
orthogonal group, 16
orthogonal projection, 23
outliers, 111
overall joint influence, 139

P

percentile, 71
polynomial model, 105
population, 51
population covariance, 84
population mean, 51, 56, 56, 66, 67
population parameter, 57
population variance, 58
positive-definite, 12
pre-normalized q-influential, 149
predictor, 93
principle curvatures, 38
principle directions, 38
probability density function, 53, 65, 80
probability distribution, 53
probability function, 53
probability mass function, 53

R

random sample, 51, 53
random variable, 52
range, 58
rank, 7, 161
regression, 93
regression line, 97
regression plane, 107
relative frequency histogram, 55
residual, 99, 107
residual sum of squares, 100, 103, 108
residuals, 99
response variable, 93
Ricci curvature, 163

S

sample correlation coefficient, 82
sample covariance, 82
sample mean, 51, 57, 57
sample variance, 57
sampling distribution, 57, 62, 74, 75
scalar multiplication, 3
scatter plots, 83
second fundamental form, 36
sectional curvature, 40
simple linear regression, 93
square matrix, 5
standard basis, 4
standard deviation, 58, 58
standard normal distribution, 69
statistical inference, 51
statistical model, 52
statistics, 57
sum of squares due to error, 100, 108
summary statistics, 57
symmetric, 24

T

tangent space, 10
tangent space of the graph, 30
trace, 8
translation, 16
transpose, 5

U

unbiased, 67
unbiased estimator, 60, 76

V

variance, 58, 58, 59
variance covariance matrix, 88
variation due to regression, 103
vector addition, 3
vector subspace, 6
volume of the parallelepiped, 26

W

weighted least squares, 103
weighted least squares estimates, 103
weighted least squares estimator, 110

Made in the USA
Monee, IL
16 November 2025

34961387R00105